Studies in Fuzziness and Soft Computing   290

Editor-in-Chief

Prof. Janusz Kacprzyk
Systems Research Institute
Polish Academy of Sciences
ul. Newelska 6
01-447 Warsaw
Poland
E-mail: kacprzyk@ibspan.waw.pl

For further volumes:
http://www.springer.com/series/2941

Kofi Kissi Dompere

Fuzziness and Foundations
of Exact and Inexact Sciences

 Springer

Author
Kofi Kissi Dompere
Department of Economics
Howard University
Washington, D.C.
USA

ISSN 1434-9922   e-ISSN 1860-0808
ISBN 978-3-642-44535-4   ISBN 978-3-642-31122-2 (eBook)
DOI 10.1007/978-3-642-31122-2
Springer Heidelberg New York Dordrecht London

Printed on acid-free paper

Springer is part of Springer Science+Business Media (www.springer.com)

Dedication

To those philosophers, logicians and mathematicians, past and present involved in the problem of representation of inexact information structure, the study of the problem of vagueness in both the ontological and epistemological spaces.

To the system of Pharaonic concepts of opposites;
To the Dogon people for their concept of duality;
To the Akan people for the Adinkra symbolism of thought in polarity, duality and logical continuum.

To Janusz Kacprzyk for his dedication to the editorship of the Springer Series on Fuzziness and Soft Computing in all areas of knowledge production, including the development of foundations of fuzzy processes, decision-information systems and their applications to hard and soft sciences, technology, engineering and mathematics with the view of connecting the past, present and future in the spirit of Sankofa.

In the memory of:
My mother and her tenth child, Yaw Badu
My professors Karl H. Niebyl and Norman Sun of Temple University
My colleagues: Professors S.Y. Kwack, Cleveland Chandler, Byung S. Lee, DuCarmel Bocage and Frank Davis of Howard University.

Acknowledgements

The theory of modeling has helped in gaining important insides into a number of phenomena through its mathematical and logical structure. Such insides reflect intelligent rules of decision-choice behavior in both humanistic and non-humanistic systems. The set of intelligent rules of decision-choice behavior constitutes rationality. The insides, however, have been confined to the limits set by classical paradigm composed of its mathematics and logic. These limits have led to critical examinations of the classical paradigm, particularly, in situations involving complex and humanistic systems. Here, our thanks go to Herbert Simon in bring into our attention the boundedness of classical optimal rationality in decision-choice processes. Great thanks go to Joan Robinson for her insistence to keep in mind the questions that any theory is seeking answers to. We also express an appreciation for all who have worked and still working on the theory of criteria and its relationship to the understanding of decision-choice rationality within the classical paradigm. These works, however, are in domain of Aristotelian logical inference where vagueness, ambiguities and linguistic reasoning are not incorporated as part of our reasoning process.

Mathematical and logical modeling of fuzzy phenomena is the answer to the limitations of the classical approach. It has made impressive contribution to the understanding of the foundations and limitations of classical mathematics and exact sciences, and the role that vagueness, inexactness and ambiguities play in human reason and decision-choice actions. Fuzzy paradigm is pioneering what may happen, by relaxing the classical assumptions of exactness, without asserting the final structure. Great thanks, therefore, go to all the researchers, scholars, scientists and mathematicians whore are devoted to the new and important areas of knowledge construction on the basis of fuzzy logic. This monograph has benefited from the works of different authors in classical and non-classical setting. I thank Ms. Jasmin Blackman for her editing suggestions. I express my gratitude to the editors of Studies in Fuzziness and Soft Computing and for the wisdom of Springer in publishing the series. I would also like to express my appreciation for the family of unpaid programmers, listener-supporters and the paid staff of WPFW Radio 89.3 fm in Washington DC, U.S.A. Finally, a special thank you goes to Semenawit Ghebruwubet for her encouragements and motivations.

Preface

There is a striking contrast between our ability in everyday practice to successfully recognize objects, events, properties etc., and, on the other hand, our inability to provide complete descriptions or unambiguous characterizations of them. Experience tells us that any attempt at a complete characterization of a piece of reality eventually reaches a point of uncertainty as to the adequacy and /or applicability of identification criteria suggested Jiří Bečvář *[R19.62, p. 1] [R19.3, p. 1].*

In their attempt to know more about "knowledge," Dewey and Bentley have regarded knowing as an aspect of human behavior. No inner knower, or mind, or soul that does the knowing is assumed. They take man as they find him behaving in his cosmos or universe, never apart from it. Likewise they take whatever is known in the cosmos or universe as they find it, never in isolation from it [R16.14, p.269].

An exact description of any real physical situation is virtually impossible. This is a fact we have had to accept and adjust to. As a result, one of the major problems in description (essential to communication, decision making, and, in a broader sense, to any human activity) is to reduce the necessary imprecision to a level of relative importance. We must balance the needs for exactness and simplicity, and reduce complexity without oversimplification in order to match the level of detail at each step with the problem we face.[R4.4, p.149].

In our quest for precision, we have attempted to fit the real world to mathematical models that make no provision for fuzziness. We have tried to describe the laws governing the behavior of humans, both singly and in group, in mathematical terms similar to those employed in the analysis of inanimate systems. This, in my view, has been and will continue to be a misdirected effort, comparable to our long-forgotten searches for the perpetuum mobile and the philosopher's stone.[R4.49, p. ix].

Certainly, objective knowledge must be the purpose of human evolution, but a permanent modesty must be present in the researcher and engineer; we know nothing of reality other than through our models, our representations, our more or less true laws, our acceptable approximations in the state of our knowledge. And the model of something for one is not exactly the same model of this thing for another; the formula may remain the same, but the interpretation may be different. The universe is perceived with the aid of models that are indeed perfecting themselves through embodying one in other, at least until some revolution in ideas appears no longer permitting a correct embodiment.[R4.49, p. xi].

The monograph presents fuzzy rational foundations of the structure of exact and inexact sciences as seen over the epistemological space which is distinguished from the ontological space. It examines the problem of constructing the epistemological space and how this epistemological space is related to uncertainties and the paradigms of thought, exact science and inexact science and knowledge development in general. The epistemological space is argued to be general to the knowledge-production enterprise. It also provides the relational foundation to examine the demarcation problem in the knowledge space. In this age of modern science and technology, with rapid transformations in nature and society, the world needs all her global thinkers, scholars and non-scholars where no segment of the global community must sit as spectator. The history of knowledge development, scientific progress and technological advancements suggests to us that these human events are not spectator sports where others play the intellectual game and others sit at the sidelines and watch. The path of the progresses of science, technology, engineering and knowledge production is the path of participatory democracy. Its development depends on the effective mobilization and organization of human and non-human resources through the casting of relevant institutions to optimize social creativity, imagination, thinking and innovation of cognitive agents for global information-knowledge production in terms of science, technology, engineering and mathematics. The study of institution and institutional casting for the effective mobilizations and organizations of human and non-human resources are the responsibility of the social science and the corresponding social technology that houses the creative imaginations and the will of people to actualize the potential through the process of creative destructions in substitution-transformation dynamics under the principle of cost-benefit rationality.

The rapid advancements in modern science require the understanding of paradigms as vehicles to process information into knowledge by cognitive agents. These vehicles are under continual innovations and improvements as new knowledge becomes actualized and outdate knowledge fades into the potential. From this viewpoint, the roles of fuzzy rationality and the classical rationality in these examinations are presented in relation to their corresponding paradigm and laws of thought in cgnition. The driving force of the discussions is the nature of the information about the epistemic elements that connects the cognitive relational structures of the epistemological space to the elements in the ontological space for knowing. The knowing action is undertaken by decision-choice agents who must process information to derive exact-inexact or true-false conclusions. The information processing is done with a paradigm and corresponding laws of thought that constitute the epistemic machine for input-output transformation from information to knowledge, from epistemic potential to the epistemic actual. The kind of the paradigm selected depends on the nature of the information structure that is taken as an input for the thought processing. Generally, the information flowing from the ontological space is pure, but it acquires a character of defectiveness at the epistemological space from the simple principles of acquaintances and the limitations of cognitive agents operating in the constructed epistemological space in attempts to cognitively reach the ontological space which is not directly accessible. How then do we arrive and claim exactness in our knowledge-production system? How do the information structures affect the paradigms of thought; and how do both of them affect the outcome of the knowledge-production activities?

In this respect and in relation to the questions asked, there are four elements in the category of knowing which may be called the epistemic system. They are the epistemological space, the ontological space, the information space and the processing space. This epistemic system is a knowing structure for all areas of knowledge production, and these areas of knowledge production whether they are called exact or inexact science, must work with the *defective information structure* generated by acquaintances and perceptions even if it is enhanced by instrumentations. The problem associated with the principle of this acquaintance is reflected upon by Frank Knight who states:

> *We perceive the world before we react to it and we react not to what we perceive, but always to what we infer.... The universal form of conscious behavior is thus action designed to change a future situation inferred from a present one. It involves perception and, in addition, twofold inference. We must infer what the future situation would have been without our inference, and what change will be wrought in it by our action. Fortunately or unfortunately none of these processes is infallible, or indeed over accurate and complete. We do not perceive the present as it is and in its totality, nor do we infer the future from the present with any high degree of dependability, nor yet do we accurately know the consequences of our own actions* [R16.16, pp. 201-202].

The problem of concern is to examine the conditions that allow claims of exact science, inexact science, and inter-supportive systems, and to make a convincing argument for the study and use of the fuzzy paradigm. There is a revolution taking place in the contemporary system of knowledge production. This revolution is making clearer and clearer the relationships among matter, energy and information and how information is expanding the field of human understanding and knowing of natural and social events and processes, and the relationship between them. The same information is opening new problems of human ignorance that presents challenges of knowing. Some aspects of the understanding and knowing have come to be classified as exact and inexact sciences in the knowledge-production space. This classification requires our ability to place each one in the proper location in the epistemological space and relate them to the epistemic process and the corresponding paradigm of thought. In the process of placement of the classificatory elements and the corresponding category of thought two epistemological spaces are presented. They are exact epistemological space and inexact epistemological space. The reflection of Bertrand Russell (from his book *The Wit and Wisdom*) is useful in understanding the relative structure between the exact and inexact epistemological spaces as they relate to exact and inexact sciences.

> *Although this may seem a paradox, all exact science is dominated by the idea of approximation...When a man tells you that he knows the exact truth about anything, you are safe in inferring that he is an inexact man. Every careful measurement in science is always given with the probable error, which is a technical term, conveying a precise meaning....A scientific opinion is one which there is some reason to belief true; an unscientific opinion is one which is held for reason other than its probable truth....The power of using abstractions is the essence of intellect, and with every increase in abstraction the intellectual triumphs of science are enhanced.*
>
> *What science does, in fact, is to select the simplest formula that will fit the facts. But this, quite obviously is merely a methodological precept, not a law of Nature. If the simplest formula ceases, after a time, to be applicable, the simplest formula that remains is selected, and science has no sense that an axiom has been falsified.*

The exact epistemological space is artificially created as a classical embedding that is induced by a set of assumptions, and corresponding to it is the classical paradigm of thought with a corresponding logic and mathematics. The inexact epistemological space is naturally generated and induced by human cognitive limitations, and corresponding to it is the fuzzy paradigm of thought with its logic and mathematics. The inexact epistemological space contains the exact epistemological space while the fuzzy paradigm forms a covering for the classical paradigm. It is here that Russell's reflection becomes helpful in understanding the framework of the fuzzy paradigm. This framework is supported by Brouwer and the intuitionists' positions on logic and mathematics in the knowledge production. The general conclusion is that the conditions of the fuzzy paradigm, with its laws of thought and mathematics, presents a methodological unity of exact and inexact sciences, where every zone of thought has fuzzy covering. Zone one acts as the primary category from which all other zones are derived. In this way, inexactness constitutes a primary category from which degrees of exactness are derived, and uncertainty constitutes the primary category from which degrees of certainty are also derived in the knowledge-production process. A further conclusion is that in all knowledge areas, only certainty-value equivalences and exact-value equivalences can be derived and these are the best that cognitive agents can hope for. The exact-value and certainty-value equivalences establish the knowledge distance which is the difference between ontological and epistemic elements. The knowledge distance may be measured by the fuzzy-stochastic conditionality through the membership characteristic functions that are established over inexact probabilities. Within the framework of fuzzy paradigm, every epistemic reality is always conditional waiting for an epistemic tuning; final reality resides in the ontology, not in the epistemology. Through the epistemological information, matter is conceived in terms of its characteristic sets. From the conception of matter, energy field is projected; and from the energy field, forces of quantitative and qualitative motions are defined for the transformations from one element of matter to another element of matter in society and nature. The epistemic realities define temporary transitional points that are established by decision-choice actions by cognitive agents on the basis of initial conditions, paradigm of thought and epistemic transversality conditions over penumbral regions of epistemic decisions.

The monograph is set in the following chapter sequence. In Chapter 1, we examine the characteristics of exact science and how these characteristics are used to define exactness of science in relation to the classical paradigm of thought. The central driving force in the discussion is centered on information and the manner in which information is obtained and processed. The exactness of information is related to human cognitive deformity. The cognitive deformity is related to the epistemological exactness relative to ontological exactness and then connected to knowledge distance and how the knowledge distance is related to incompleteness, vagueness and ambiguities in the information connector and processor. The incompleteness, vagueness and ambiguities give rise to epistemic questions on quantity-quality relations on science and knowledge processes and then connected to the problems of claims of exactness of science. The discussions are related to some important works of Max Planck, Amo, Max Black, Zadeh, Russell and others where geometry of epistemic unity is provided to create a framework of the role of decision-choice rationality and permissible limits to claim exactness in the knowledge search and what the knower knows.

Chapter 2 is used to discuss the problem of the exactness of the derived categories from the primary category of science in relation to the epistemological information structure where the relational dynamics of matter-energy-information are used to illustrate both the point of entry and the point of departure from the viewpoint of exactness in science. The concepts of fuzzy and stochastic residuals are introduced, explicated and related to the concept of irreducible core of inexactness of science where Max Planck's concern is made explicit with quantity, quality and time. The exactness of the classical mathematics is examined and the problem of information representation by exact symbolism is discussed.

The approaches by Ludwig in relation to inexactness in physics, and those of Max Black to Mathematics are reflected on and then related to the concepts of probabilistic and fuzzy tuning. The essential characteristics of the classical and fuzzy paradigms are presented in a comparative mode and related to exact and inexact epistemological space as an explanation to E. Grafe's concerns as the conclusion of the chapter.

In Chapter 3, the concepts of exact and inexact sciences are related to Max Planck's concept of a scientific world picture and then to the derived categories and the primary category over the epistemological space. The problems of exact-inexact duality are then discussed and related to the concept and process of continual tuning to improve the validity of epistemic claims in the framework of fuzzy-stochastic spaces that relate to the evolving world pictures or the derived categories of knowledge. The path of the knowledge-production process through the possibility and probability spaces to the space of epistemic actual are then related to the evolving inexact scientific world picture over the time points. A discussion of the path as capturing the history of exactness of science as an enveloping of epistemic equilibria is provided and connected to the problem of scientific discovery or knowledge discovery.

The conditions of questionable claims of exactness and certainty in science are discussed in relation to decision-choice rationality through the paradigms of thought over the epistemological space. The concept of *fuzzy conditionality* is introduced and related to Max Black's concept of *qualification of degree of consistency* and Günther Ludwig's concept of *qualification of degree of uniformity*. The chapter is concluded with relational structure of empirical and axiomatic conditions in the establishment of the primary category in the knowledge-production process.

In Chapter 4, a presentation is made of the organizational structure of the knowledge-production process and as a defense of inexact science. The knowledge-production system as we have presented is a decision-choice system involving individuals and the collective under an organizational structure which is continually changing under information-knowledge conversions. The organization of knowledge-production is part of the social institutional configuration which is composed of legal, political and economic institutions that present control systems of social activities of the knowledge-search agents. The organization and the corresponding control systems are the same for exact and inexact sciences. The study and the knowledge production of this mega organization is under social sciences which is claimed to be inexact by the supporters of exact sciences. The knowledge of this inexact science sets the parameters of the exact science, it is argued. In this respect, it is argued that the degrees of exactness of exact science are constrained by the degrees of inexactness of inexact science, both of which are related to the nature of defective information structure over the epistemological space.

The defective information structure is a phenomenon in all knowledge sectors and hence exactness and inexactness are factors of all aspects of the knowledge-production process where the cognitive agents must work with this information structure under some decision-choice criteria. The criteria of partition of the knowledge space are not in the content of the subject but in the nature of simplicity, methods and technique of the knowledge search that lead us to discuss the unified epistemic methods of the knowledge production. From the criteria of partition, the exactness and certainty of inexact science are then discussed with a list of different and similar characteristics of the exact and inexact sciences. The list leads us to the discussion of the concepts of quality, quantity and time which are then related to exact and inexact symbolisms and the appropriate logic of reasoning. The chapter is concluded with reflections on the basic operators of classical and fuzzy numbers and the possible epistemic differences and similarities for reasoning and the construction of computable systems that relate to duality, opposites and continuum principles.

Chapter 5 is used to discuss, in essential details, the relational structure of duality, and continuum in the fuzzy paradigm and the corresponding laws of thought and how they relate to exactness in inexact science and inexactness in exact science to demonstrate that the claim of exactness in any area of knowledge production is decision-choice determined with conditionality. The analytical structure is developed through the organization of the knowledge-production process. The nature of the universal principles of ontology and epistemology as seen in terms of the ontological information and epistemological information are advanced.

The role of language and sources and nature of vagueness are discussed in relation to defective information structure over the epistemological space where special emphasis is placed on the qualitative component of the information structure and then related to the need for the use of the fuzzy paradigm in explication of the concepts of exactness and inexactness in the knowledge-production process including science. The explications and the uses of dualism, duality and unity in scientific cognition are presented and linked to a cognitive geometry and a fuzzy algebra of thought. From the fuzzy algebra of thought, a distinction is made between the classical and fuzzy paradigms where the fuzzy rationality in knowledge production is related to the cost-benefit rationality in the knowing process. The conditions of fuzzy theory of truth are pointed out leading to the comparison and discussion on the theories of truth.

Chapter 6 concludes the monograph with discussions on the zones of thought where reflections are made on the classical and fuzzy theories of truth and exactness which are then related to the concerns of Russell, Brouwer and Black, and the intuitionist mathematics. Four zones are identified. Each zone corresponds to a particular information structure and its symbolic representation that define the conditions for the laws of thought that are appropriate for processing the information into knowledge. The partitioning of the epistemological information space allows for the discussions of zones of unity of exact and inexact sciences in the knowledge-production enterprise. In the process, a discussion is made on simplicity-complexity duality in the knowledge production and how it is related to truth-false verification and exact-inexact determination.

Zone I is identified with complete uncertainty of information vagueness and incompleteness with *fuzzy-random* (*random-fuzzy*) variable as its information-representation symbolism where the space of analytical work in terms of logic and mathematics is the *fuzzy-stochastic topological space*. Zone II is identified with exact and incomplete information where there is a partial uncertainty due to information incompleteness. The symbolic representation of the information structure is *exact-random variable* where the space of analytical work in terms of logic and mathematics is the *exact stochastic (probability) topological space*. Zone III is identified with exact and complete information where there is no uncertainty. The symbolic representation of the information structure is *exact-classical variable* where the space of analytical work in terms of logic and mathematics is the *ordinary classical (non-fuzzy and non-stochastic) topological* space. Zone IV is identified with an information structure that is vague but full with only fuzzy uncertainty. The symbolic representation of the information structure is *non-stochastic fuzzy variable* where the space of analytical work in terms of logic and mathematics is the *non-stochastic fuzzy topological space.*

The reflections lead us to the conclusion that the fuzzy paradigm applies to all the four zones. The classical paradigm with its laws of thought and mathematics applies to zones II and III where there is no ambiguities and vagueness in the information sets and where the information structure is complete or incomplete. The general conclusion is that the conditions of the fuzzy paradigm with its laws of thought and mathematics present a methodological unity of exact and inexact sciences where every zone of thought has a fuzzy covering. Zone one acts as the primary category from which all other zones are derived. In this way, inexactness constitutes the primary category from which exactness is derived and uncertainty constitutes the primary category from which certainty is derived in the knowledge-production process. In all knowledge areas, only certainty-value equivalences and exact-value equivalences with conditionalities can be derived; and these are the best that cognitive agents can hope for. The chapter is concluded with some classes of fuzzy numbers that are relevant for reasoning in vagueness, categories, opposites, duality, polarity, continuum and contradictions. It would have been useful if a glossary of terms and a list of principal symbols were added to assist the readers. This was not done since the terms and symbols are well defined in the monograph. The hope is that this monograph will bring a fresh understanding of the differences and similarities between the exact and inexact sciences from the position of the fuzzy paradigm.

Prologue

The development of the contemporary information-knowledge system and the fast moving technological know-how in general human endeavors demand a continual examination and reexamination of information and knowledge and their impact on quality of human actions and social systems. This development places on us a new framework in examining our concepts of epistemology and ontology. Decision-choice agents function as cognitive agents over the space of knowledge but not simply the information space before undertaking decision-choice action in all spaces of human endeavor. The idea of knowledge production as an output from information processing suggests that information is not knowledge; it has to be refined to a point where knowledge can be claimed with a justification. Knowledge depends on cognitive capacity and limitations as reflected on the epistemic channels of cognition. The manner in which these epistemic channels of knowing come to us is where we know turn our attention.

I: On the Epistemological Space and Science

The epistemological space is a cognitive construct. It is the field of the knowledge-production game which is available to all cognitive agents. The objective in constructing the epistemological space may be seen in terms of two epistemic actions of explanation and prescription. At the level of explanatory epistemic system, it creates an environment that contains the guidelines to knowledge discovery involving *what there is,* the knowability of *what there is* and *what does the knower know* in terms of explanation of behavior? At the level of prescriptive system, it creates an environment that contains the guidelines to knowledge discovery involving what there is (the actual) and what ought to be (the potential) and how to actualize the potential. The epistemological space is a construct of information, cognitive processor, decision-choice modules, quality-control and knowledge.

The decision-choice module is the epistemic core of the epistemological space. It contains cognitive agents who are also decision-choice agents whose actions define the information structure, the information processor, quality control and knowledge acceptance into the epistemic reality. In this way, the cognitive agents provide the dynamic force of changes in quality and quantity of knowledge output as well as a force for internal transformation of the epistemological space in such a way that the knowledge-production system acquires the properties of self-correction, self-transformation and self containment where changes are always internally self-induced. The epistemological space is part of the knowledge production system. By equating the cognitive agents with the decision-choice agents we have introduced the notion that the epistemological space and the corresponding information-knowledge production system are social system constructs. The knowledge-production system is seen as a subset of the social system while the outcomes of the knowledge-production shape the quantity and quality of the social system through the decision-choice activities and the use of knowledge by the decision-choice agents in restructuring the social system when the outcomes deviate from the desired. The disparity, between outcome and expected, triggers the knowledge controllers to influence changes in the structure of the epistemological space and the knowledge-production system for correction. The social system is the primary category of cognitive construct while the knowledge-production system is a derived category by categorial transformation.

The utility of the construction of the epistemological space is to create connecting paths between the cognitive agents and the ontological space in order to examine the knowing processes between epistemic elements and the ontological elements. In this way, there is a claim of an element of the epistemic reality when there is a justified isomorphism between an epistemic element and any ontological element. The epistemological space does not contain knowledge, neither does it contain reality. It is simply, the working space of cognitive agents and knowledge seekers working with their tools and information to build the universal knowledge house of humanity with continual room expansion and refinements where each room is a subsector with epistemic links. The knowledge house constitutes the epistemic reality at any decision-choice time or period. Two interdependent technological processes evolve from the epistemic reality. They are social technological process and it progress, and physical technological process and its progress at the service of humanity.

II: Reflections on the Epistemology, Exactness –Inexactness Duality and the Demarcation Problem in the Knowledge-Production Process

The construct of the epistemological space and the path to the epistemic reality require a number of building blocks given the ontological information flows. These building blocks include information-structure development, language development, category formation on the basis of the language, concepts, definition, explication, measurements, and development of toolbox of information processing within the language. The development of the information structure creates the input requirement for potential reduction of human ignorance. The information structure may have empirical or axiomatic basis in specifying its existence. Language is a cognitive instrument for information representation. Item classification by naming allows the development of concepts and category formation with the information representation that is either partially or fully understood by cognitive agents working in the same language regime. Concept definitions and explications affirm the clarity of communications in the language of thought. Measurement is geared to the distinction between the qualitative and quantitative concepts. The toolbox of information processing involves the logical manipulation of the information contents of concepts and events to derive a knowledge content of the information to reduce ignorance.

The initialization of the journey over the epistemological space rests on the descriptive characteristics of the information input which involves the creation of either an *axiomatic foundation* or an *empirical foundation* or the combination of both of the information. The empirical foundation is drawn from sensory experiences and conceptual descriptions. Generally, in the scientific workplace, empirical information originates from sense-impressions or sense data which are not the reality itself but symbolic representation, such as formal and informal words. Its goal is the provision of input to generate understanding through concepts and concept combinations. The concepts are reflections of properties of elements, states and processes which constitute the social and natural phenomena. In this respect, one may view the basic problem of the theory of knowledge production as the relations of concepts to sensory data.

Here, experience through acquaintance presents us with the origin of concepts which are their abstractions while language as symbolic system presents us with the mode of representation and the role of concepts is to order human experiences. If empirical information originates from sense-impressions or sense data through acquaintances then where does axiomatic information originate? The axiomatic information like the empirical information is not the reality itself but symbolic representation such as formal and informal words. Alternatively, what is the source of axiomatic information and which one can we claim to be the exact and valid input for the information-knowledge processing enterprise?

Those who subscribe to the validity of empirical information, also claim that any knowledge about a reality has an experiential basis and terminates in that experience and that pure logical thinking is a tool that provides us with no knowledge about the experiential space. It helps the establishment of the relational structure of experience. Given the information foundation as input into the knowledge-production process the creative principle is the use of logical and mathematical constructions. The use of the combination of the two creates a theoretical framework to discover concepts and laws that connect them to generate understanding of natural and social phenomena as well as to build knowledge system. This knowledge system constitutes the epistemic reality which is under continual self-refinement as a self-correcting system. From the position of those subscribing to empirical information, the axiomatic information provides no framework to derive knowledge except it can be shown to have an experiential basis.

From the general epistemological space, arises the demarcation problem of the knowledge production enterprise. The problem simply involves the partition of the epistemological space into subspaces of knowledge search. The partition starts with science and non-science where the science is further partitioned into exact science and inexact science. The categories of exact science and inexact science are categorized into subspaces for many reasons. The nature of the demarcations must relate to the information-structure development, language development, category formation on the basis of the language, concepts, definition, explication, measurements, and development of toolbox of information processing within the language that helps to describe the epistemological space. Every partition in the knowledge space requires a criteria which may be composite or simple, vague or sharp, qualitative or quantitative or both. There are many aspects of the demarcation problem. In this monograph, our concern is on exact and inexact sciences and the nature of the criteria that can be used to claim exactness of science and inexactness of science. The criterion must be based on all or some aspects of the building blocks of the epistemological space. Given the building blocks of the epistemological space, the demarcation problem relative to the interest in this monograph takes on two steps in problem solving. The first step is to find the set of conditions that separates science from non-science. The second step is to answer the question: What is exact science and what is inexact science given the explication of science and what conditions give them their identities as well as separate them into recognizable units?

The demarcation may be done on the basis of division of intellectual labor and specialization to induce high intellectual output. It may also be done on the basis of ideology to specify the conditions of science and non-science where the results of the former is claimed to be scientific knowledge with increasing social trust while the results of the latter is claimed to be unscientific with increasing social distrust in its use. A similar situation is encountered in the relative positions of exact and inexact sciences where the criteria of partition are also ideological in order to specify the set of conditions where the results of the former is claimed to be exact scientific knowledge with increasing social trust in its use, while the results of the latter is claimed to be inexact scientific knowledge with decreasing social trust in its use. The interesting point here is that all the building blocks of the epistemological space are inexact and the best we can do is the construction of inexact epistemological space relative to the ideal which is exactness. In this respect, no criterion can be constructed from them to claim unconditional exactness of science.

The information-structure of the epistemological space is made up of categories of epistemic elements. These elements are identified by the qualitative and quantitative dispositions to provide them with identities and concepts. The identities and concepts as seen in terms of qualitative and quantitative characteristics are connected to a defined language development. Given the quantitative and qualitative characteristics, categories of epistemic elements are formed on the basis of the language and concepts. The concepts acquire inexact information contents, each one with a multiplicity of meaning in the natural language. Any naturally linguistic word covers a usage space that is shared by different knowledge sectors in such a way that the word by itself is vague in information content. It is here that definitions of words and concepts demand restricted domain leading to the theory of definitions over the epistemological space to specify the applicable area of thought. These defined terms and concepts contain elements of vagueness that is particular to the subarea of knowledge search.

The vague content of information in subareas of thought and epistemic search demands further definition called the explication and corresponding to it the theory of explication. The role of definition is to reduce vagueness and ambiguities and the theory of definition is to provide an algorithm to accomplish this task. The role of explication is to further reduce the vagueness and ambiguities in specific areas of thought to refine the information content of the word and concept. In general, the reductions in vagueness and ambiguities are to reduce inexactness inherent in the epistemological space so as to increase the degree of exactness for the human thought development. The defined and explicated concepts retain some relatively irreducible core of vagueness due to the presence of qualitative characteristics contained in them. The simultaneous presence, of qualitative and quantitative dispositions in refined concepts, does not easily allow manipulability, combinations and aggregations within the linguistic structure. The concepts are then examined for measurement possibilities to get rid of the qualitative complexity so as to further reduce the vagueness and ambiguities in the epistemological space. This leads to the theory of measurements that takes on different complexities depending on the particular area of thought and the nature of qualitative characteristics.

The measurement to rid the linguistic variables of the qualitative characteristics allows the creation of what is referred to as scientific information by reducing vagueness to nothingness. This allows the use of exact symbolism for information representation. The process, thus artificially restrict the epistemological space to its quantitative characteristics at the expense of qualitative disposition of epistemic elements to create an exact epistemological space. The exact epistemological space by a decision-choice construct that allows the use of the exact logical and mathematical construct and reasoning, therefore, is a subspace of the inexact epistemological space that is natural to cognition. In this way, scientific work is restricted to the processing of quantitative information to produce aspects of epistemic reality. It is this problem of exact and accurate symbolism with exact information representation that occupied Ludwig Wittgenstein in his work, *Tractatus Logico-Philosophicus*, to find the conditions for accurate and exact symbolism in which concepts have definiteness and in which mathematical objects acquire absoluteness in information representation. When the measurement is exact to produce exact information then the area of work is exact science. In this exact scientific workplace, the classical paradigm with its logic and mathematics, satisfying the Aristotelian principle of non-contradiction and excluded middle, becomes a useful tool for information processing. The distinction of mathematical object from epistemic object is seen in terms of the content of qualitative characteristics.

Here, a problem arises regarding the exactness of the measurement and the usefulness of exact symbolism to derive exact thought with relevance, especially when the complexity of the phenomenon increases. Exactness and relevance are seen to diverge as the complexity of the phenomenon increases. This is another way of conceptualizing the Zadeh *incompatibility Principle:*

> *Stated informally, the essence of the principle is that as the complexity of a system increases, our ability to make precise and yet significant statements about its behavior diminishes until a threshold is reached beyond which precision and significance (or relevance) become almost mutually exclusive characteristics. It is in this sense that precise quantitative analyses of behavior of humanistic systems are not likely to have*

> *much relevance to the real-world societal, political, economic*
> *and other types of problems which involve humans either as*
> *individuals or in groups.* [R4.69, p.9]

The reason for this incompatibility is that part-to-part interactions and differentiations in the complex phenomenon lead to increasing qualitative disposition in the defining characteristics which become more and more difficult to quantify. We must, however, keep in mind that in the quantity-quality duality, there is a continuum such that every qualitative disposition has a corresponding quantitative disposition and vice versa. The meta-theoretic implication, here, is simple in that every qualitative characteristic is a potential quantitative characteristic. Similarly, every quantitative characteristic is a derivative of a previously existing qualitative characteristic.

III: Quality-Quantity Duality, Fuzziness and Science

Both qualitative and quantitative information characteristics are related to human concepts of state and change. Generally, there is an ideological fixity in the knowledge-production enterprise where scientists are not satisfied with their knowledge and analysis until they have succeeded in transforming qualitative disposition into quantitative disposition in the process of understanding change. This presents an increasing preoccupation with quantitative dispositions of state and change. The state of the phenomenon is viewed solely in quantitative disposition that has constant qualitative disposition. The result is a complete neglect of natural and social transformations involving the understanding of qualitative motions and simultaneous changes of quality and quantity where both the qualitative and quantitative motions are locked in our experience to create complex information system for processing into knowledge. This preoccupation is reflected by E.C. Harwood

> *Now quantitative change is measured change; that is, we*
> *recognize that change has occurred by measurement, however*
> *crude that measurement may be. In some instances the*
> *measurement may be so crude that the changes are stated in*
> *such inexact terms as greater or less, higher or low, faster or*
> *slower, longer or shorter, high speed has been retracted, rising*
> *prices have changed to falling prices, what used to require a*
> *week can now be done in less time, these are examples of rough*
> *measurements of quantitative changes* [R8.26, p.4]

The preoccupation on quantitative disposition has negative and positive effects in the knowledge-production process. The positive effect is that there are continual intellectual activities of finding methods and ways to quantify the qualities through some forms of measurement in every area of knowledge search. Additional positive effect is reflected in communication if measurement succeeds, where the quantitative disposition helps to reduce disagreements in knowledge claims. The result is the continual refinement and improvement of theory of measurements and corresponding technologies of measurements. The negative effect is that there are tendencies to neglect the essential elements of qualities that generate identities of elements in natural and social transformations and where attempts are made to do away with the study of a phenomenon if the qualitative dispositions are overwhelming and can not be transformed into quantitative dispositions. The quantitative disposition characterized in classical exact symbolism has come to define scientific ideology that allows us to use the classical paradigm and its corresponding mathematics and logic for the information-knowledge transformation process over the epistemological space.

A number of knowledge seekers and cognitive agents are under the influence of this classical ideology of knowledge production. Armed with this ideology, they have come to claim that truly creative principle in theories in various areas of knowledge is *mathematical constructs* to represent the needed epistemic frames. The supporting reasoning of this claim of creativity is that mathematical constructs allow the cognitive agents to find exact concepts and connectors to provide a better understanding than without them. Here we must keep in mind that words of languages are just as symbolic as any symbolic system. Any symbol can be thought of as a word. It is the choice of these symbols and connectors that generate ideas from the epistemic processing technology to create a picture. The words may be thought of as colors that may be selected and coordinated to establish pictures through the painting process. These colors of the ordinary language have vague boundaries where one color fades slowly into another color with shared indistinguishable boundaries that create penumbral regions in thought and application. The notion of exact symbolism is to sharpen the individual colors and demarcate the boundaries for distinction. In so doing, the penumbral regions of knowledge become lost where the knowledge search is concentrated on exact boundaries. The question within cost-benefit analysis involves with what we gain and what we loose in the knowledge-production process over the epistemological space. This again brings us to the question of exactness and relevance. The opportunity cost of exactness is the loss of knowledge in the penumbral regions of search and the size of the loss is in proportion to the size of the aggregate penumbral region in the epistemic process.

Another question tends to arise. What mathematical concepts in relation to symbolism are appropriate to represent the needed theory in order to provide us with both exactness and relevance in analyses and understanding? Here, one may view any theory in any knowledge area as a rational construct to reduce the conceptual system that is directed to establish an epistemic reality where the size of the conceptual system is minimized relative to the available information structure. In the knowledge-production process, we must solve the problem of minimal conceptual system subject to relevance or maximize relevance subject to appropriate mathematical concepts and symbolism.

The general practice and our current experience in the mathematical representations of theoretical forms have been that kind of mathematics that fits into the structure of the classical paradigm with its exact symbolism, and laws of thought that operate in logical dualism in which the principles of excluded middle and non-acceptance of contradiction apply. This process of information-knowledge transformation is being advanced to all areas of knowledge production wherever possible to obtain some level of elegance and social respectability whether it is useful or not in terms of relevance and exactness. This logical dualism and the classical mathematics are restrictive in terms of actual and possible epistemic items that may be processed in the epistemological space. In the classical framework of logical dualism, the general direction is to get rid of quality, vagueness and inexactness and basically concentrate on exactness and quantitative phenomenon. This approach is to simply do away with the trouble and nagging problem of qualitative disposition and subjectivity in the scientific progress. In this classical approach, the cognitive agents are basically externalized from the epistemological space as an active participant where subjectivity in judgment and action is factored out. The active role of the cognitive agents is transformed into a passive role where establish rules of thought are strictly followed.

The approach of the fuzzy paradigm is to find conditions that will help to develop mathematics of vagueness and inexactness that will deal with the totality of inexact epistemological space which is characterized by defective information structure. The fuzzy paradigm is not directed to get rid of the problem of vagueness but to confront it by finding the conditions that will allow the initial construct of vague symbolism to represent a defective information structure where such symbolism will allow the integration of cognitive agents as active and internal part of the knowledge-production process through decision-choice actions. The vague symbolism generates quality-quantity logically controllable fuzzy variables for inexact information representation and helps to construct and analyze fuzzy processes in state and over time. In this way, the information-knowledge production system retains its natural state of self-creation, self-generating, self-correcting and self-exiting.

The next step in the fuzzy paradigm is the designing of mathematical and logical structures that will establish operations on the fuzzy variables where such fuzzy variables are viewed as fuzzy sets, fuzzy numbers, fuzzy quantities, fuzzy relational structures and fuzzy graphs. The second step is based on the principles of opposite and duality with logical continuum and where every fuzzy variable contains its dual in an interdependent mode to establish its identity in the areas of epistemic works. The uses of steps one and two produce fuzzy logic and mathematics with fuzzy operators. They enable inexact concepts and laws connecting them to provide us with structures to understand the phenomena in inexact epistemological space and the penumbral regions of human actions through the epistemic processing of fuzzy quantities.

The development of vague symbolism and the fuzzy paradigm extend our abilities to mathematically deal with the subject areas of economics, medical sciences, sociology, complex system science, synergetics, energetics, political science, systemicity, organicity, informatics, computer science formal languages and many emerging fields of science. The sequential process involves fuzzification to obtain fuzzy epistemic variables (but not exact epistemic variables) that allow inexactness to be expressed as a mathematical fuzzy variable on which the fuzzy mathematical and logical reasoning is applied. The results of this mathematical and logical operations lead to a system of defuzzifications to obtain exact-value equivalences with fuzzy conditionality through decision-choice actions.

It is useful to note the developmental foundations of the classical paradigm and its information-knowledge process and the fuzzy paradigm and its information-knowledge process. The approach of the classical paradigm is a search for conditions that will allow inexact information to be represented by exact information to generate an exact epistemological space with exact symbolism. Given the exact symbolism, the structure of exact laws of thought and the corresponding mathematical system is erected. The approach of the fuzzy paradigm, on the other hand, is a search for conditions that will allow inexact information to be represented by inexact symbolism as it exists to retain the natural conditions of the inexact epistemological space. Given the inexact symbolism, the structure of inexact laws of thought and the corresponding mathematical system is erected through fuzzification-defuzzifisation processes. The application of the fuzzy paradigm with the corresponding mathematics and logic in the general inexact epistemological space through fuzzification-defuzzification processes lead to the derivation of general exact-value and certainty-value equivalences with fuzzy-stochastic conditionality. Armed with the fuzzy paradigm, it may be claimed that truly creative principle in theories in various areas of knowledge is *fuzzy mathematical constructs* to represent the needed fuzzy epistemic frames. These fuzzy frames have built into them the potential to assist in general systems engineering in order to deal more efficiently with the defective information structures that may be generated by system's complexity in knowledge areas such as collective decisions, stochastic and dynamic game theories, nonlinear dynamical systems, evolution-adaptation systems, self-exited and self-correction systems, non-self-exited systems, human and non-human network systems, and pattern recognitions and formations. In all these cases, changes in complexity are, in the final analysis, attributable to changes in qualitative dispositions which are the resultants of changes in the forces in the positive-negative relational structures. The supporting reason for this claim of creativity is that fuzzy mathematical constructs allow the cognitive agents to find fuzzy concepts and connectors to provide a better understanding than without them over the inexact epistemological space where increasing epistemic access is made available for abstract representations of simple and complex systems. The general methodological gains are captured by the following statement.

> *Not only is fuzzy sets theory providing the numerical tools to solve ill-posed problems, but it also is providing the specialists in various fields with a basic conceptual framework in which to formulate problems in an enlightening and perceptive fashion. Such methods could open many new frontiers in psychology, sociology, political sciences, philosophy, economics, operations research, management science and other fields, and a basis for the design of systems far superior in artificial intelligence to those we can conceive today* [R4.69, p.10].

From the fuzzy epistemic models of reality, the classical models can be abstracted through the principle of fuzzy decomposition where the methods and conditions of decomposition are made explicit. Fuzzy model of reality is an epistemic exact reality with fuzzy-stochastic conditionality. Classical model of reality is an epistemic reality with an exact-stochastic conditionality. The fuzzy-stochastic conditionality defines a grater area of riskiness than exact-stochastic conditionality. It is derived from a larger system of uncertainty than the classical exact-stochastic uncertainty.

IV: General Reflections

The greatest contribution that the current intellectuals can make to the progress of our global intellectual heritage is to avoid the dangers imposed on us by intellectual disunity, and continually search for the path of unity of sciences and the knowledge subsystems. This contribution will define our methodological greatness as an important example to the divided knowledge sectors without epistemic connecting cords. For this great contribution to be realized as well as indestructible, it must be built not on sectorial fears, envy and suspicion, and at the expense of other sectors but grounded on hope, trust and epistemic strength and generality that are directed to integrated knowledge rooms in the same house of knowledge that is directed to the good of mankind without the evils that are inherent in knowing-ignorant duality within the good-evil polarity. The monograph is motivated by this epilogue to show that exact and inexact sciences have common roots that relate to the constructed laws of thought on the basis of the assumed information structure.

The use of the classical paradigm, in human epistemic actions over the general epistemological space for knowledge production, is to assume an exact information structure that allows an exact symbolism to be constructed in the information-knowledge process. From the exact symbolism, exact laws of thought are designed for processing either the exact axiomatic or exact empirical information it into an exact rigid determination of knowledge. The system of such knowledge production is then called exact science. In this way, the question about the existence of scientific exactness is answered by assumption and exact science is then characterized and defined by its methods and the relations that are implicitly embedded in the initial assumptions of information exactness. A science is exact because we say so. In the classical system of thought, all spaces are either exact and non-stochastic, or exact and stochastic. In this way, our knowledge production is restricted into either classical exact non-stochastic topological space or classical exact stochastic topological space.

The use of the fuzzy paradigm, in human epistemic actions, for knowledge production over the general epistemological space, is to take the information as initially defective, develop the corresponding fuzzy (inexact) symbolism in representation of the information and then design inexact laws of thought for processing it into inexact flexible determination of knowledge to produce exact science with fuzzy conditionality. In this way, the existence of scientific exactness is answered by decision-choice action with allowance for inexactness. Exact science in this way, is characterized and defined by decision-choice action and human intellect on the basis of conditions that are explicitly embedded in the initial conditions of information inexactness. Exactness, in this way, is an abstraction from inexactness where both of them exist in logical duality and continuum, and where every claim of exactness has a corresponding support of inexactness. The fuzzy paradigm ensures the practice of the principle of methodological doubt, activate the intellect and liberate cognitive agents from the slavery of mental habits without innovation. It is through the practice of this methodological doubt, supported by cognitive imaginations and innovations, which functions through information-decision-choice processes, where new epistemic frames are created to enlighten the paths to knowledge discoveries as cognitive agents transverse over the penumbral regions of the decision-choice claims, moving from inexactness to exactness in degrees with the never-ending knowledge-production system. In the fuzzy frame, the central core of education and teaching, as may be abstracted from the conditions of the fuzzy paradigm, should center on the development of thinking to emancipate minds from epistemic orthodoxy that strengths the ideological boundaries which restrict human progress in knowing with judicious use of combined objectivity and subjectivity.

Contents

References

Chapter 1
Exact Science, Its Critique of Inexact Science and Rationality in Vagueness

In the theory of the knowledge square the working framework of a meta-theory of knowledge was presented to justify a claim of a universal principle of knowledge-production process that traverses from the potential space through the possibility space to the probability space and then to the space of epistemic actual. This principle is claimed to be universal to all areas of rational knowledge production. The essential revelation of the discussion is that the core of the universality principle is that the spaces of possibility and probability are defined by *defective information structure* that leads to fuzzy uncertainty and stochastic uncertainty in the information processing capacity of cognitive agents. The existence of the defective information structure is a universal principle to all areas of knowledge production ranging from the epistemic potential to the epistemic actual in the epistemological space.

The discussions were extended to the problems of definition and explication in representation of concepts by words and symbols, since these words and symbols carry a multiplicity of meanings except under specialized conditions or when they are taken as linguistic primitives. The concepts of knowledge, non-knowledge, science, non-science, exact science and inexact science are viewed as linguistic categories of human cognition in the epistemic process of trying to conceptualize the elements in the universal object set from the ontological space. The relationships between exact and inexact sciences and how they relate to the classical and fuzzy paradigms were examined where the attributes of exactness and inexactness were presented. Different dimensions of the problem of science, knowledge and paradigm are the focus of this monograph. Let us examine the critique of inexact science through the lenses of exact science and its logic.

1.1 Reflections on the Traditional Characteristics of Exact Science

In order to reflect on the exact science in relation to inexact science, it is useful to present the essential defining boundaries as currently accepted. We are made to accept that the defining characteristics of exact science are as shown in Table 1.1.1.

K.K. Dompere: Fuzziness and Found. of Exact and Inexact Sci., STUDFUZZ 290, pp. 1–14.
springerlink.com   © Springer-Verlag Berlin Heidelberg 2013

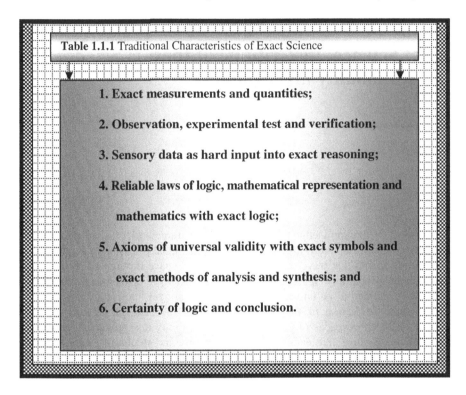

Table 1.1.1 Traditional Characteristics of Exact Science

1. **Exact measurements and quantities;**

2. **Observation, experimental test and verification;**

3. **Sensory data as hard input into exact reasoning;**

4. **Reliable laws of logic, mathematical representation and**

   mathematics with exact logic;

5. **Axioms of universal validity with exact symbols and**

   exact methods of analysis and synthesis; and

6. **Certainty of logic and conclusion.**

These characteristics cover both what is claimed to be an empirical science and an empirical theory supported by an axiomatic reasoning on one hand; and an axiomatic science and an axiomatic theory that is also supported by an empirical reasoning on the other hand. They have in common the connected claims of exactness, objectivity and the use of the classical laws of thought where the information structure is implicitly or explicitly assumed to be exact even if it is not full. It must be noted, however, that the development of the paradigm of thought will depend on whether one claims an exact information structure or inexact information structure as an input into the knowledge production process over the epistemological space. The development of the logic of information representation and the methodology for processing it begins with the nature of the information structure.

The exact science takes claim to exact information structure on the basis of which an exact symbolism and an exact paradigm of thought are developed. The specification of the boundaries of exact science takes shelter in the accepted notion that the laws of classical thought and the corresponding mathematics represented by exact symbols and mathematics, that together define the classical paradigm for the knowledge production, are unquestionably reliable in their methodological processes.   The claims of exactness and objectivity are also maintained for both methodological constructionism (forward reasoning) and reductionism (backward reasoning) in the classical domain. When the primary category, established either by empirical or axiomatic conditions, is taken to be exact and the available methodology is also exact, then, any inexact derived category, such as contradictions and paradoxes, is attributed to faulty applications of the classical laws of thought and the fault is attributed to the user's ineptness of logical reason on the basis of designed rules.

In the classical exact system, we fail to accept the deformity of the empirical nature of our claimed primary category that is due to a defective information structure as acquired by observations, experimentations and acquaintances. Furthermore, in the classical paradigm of thought, there is a refusal to accept a thought process where every conclusion rests within the true-false duality whose internal structure exists in a continuum and connected by the conversional dynamics of logical categories where the attainment of exactness is through a decision-choice process of cognitive agents. The true-values are defined only at the endpoints and this is projected to the system of knowledge acceptance in the exact sciences. The problems of logical claims of exact science and the weakness of this claim that resides in the weak structure of the exact science through its paradigm of thought are pointed out by one of its defenders, Max Planck. He states:

> *And in fact, if we take a close look and scrutinize the edifice of exact science more intently, we must very soon become aware that it has a dangerously weak point----namely, its very foundation. Its foundation is not braced, reinforced properly, in every direction, so as to enable it to withstand external strains and stresses. In other words, exact science is not built on any principle of such universal validity, and at the same time of such portentous meaning, so to fit to support the edifice properly.*
> [R8.53, p. 81]

Here, Max Plank is pointing to the problem of the defective structure and form of exact science. If the structure and form are supported by faulty cognitive pillars and maintained by a paradigm that is rigid then only a paradigm revolution can bring a needed restructuring. First, there is a problem of the exactness of information about the elements in the primary category of logical reality in the *possibility space* constructed from the *potential space* (ontological space) from which the universal object set resides. These elements contain information incompleteness and vagueness that constitute the foundation of human physical and cognitive experiments and observations as they relate to data in the process of traversing through the possibility and probability spaces to the space of epistemic actual. The characteristics of the elements in these sub-spaces are not ontological, but rather they are characteristics of the elements in the epistemological space. The ontological space contains real elements of states and processes; the epistemological space contains information about the ontological elements as conceptualized by cognitive agents for processing into epistemic reality. Secondly, there is a problem in representing vague information with exact symbols with some implied quantification.

The initialized representations of the epistemic elements must account for the observationally fundamental cognitive deformity of fuzziness and information limitation about the epistemic elements. In other words, the foundation of all knowledge and hence exact science is built on the basis of defective information structure and hence carries with it vagueness of language and ambiguities in thought that together produce *cognitive deformity* requiring the use of approximation methods to the knowledge search over the epistemological space.

Definition 1.1.1. Cognitive Deformity

Cognitive deformity is a phenomenon in the knowledge production process when there is a presence of defective information structure requiring vague symbolism in the information representation and inexact methods of information processing in the epistemological space.

This *cognitive deformity* may be associated with *fuzzy rationality* including conditions of boundedness of classical rationality as applied to the general decision-choice actions and the information-knowledge production process as a case in humans. In other words, the elements come to us as fuzzy and (or) stochastic epistemic elements in the epistemological space. It must be clear in all these discussions that the fuzziness and vagueness are not characteristics of the elements in the ontological activities [R1] [R1.9] [R1.11] [R1.17]. It is the presence of the defective information structure due to the acceptance of information incompleteness as an input into the epistemic processing machine that provides the legitimacy to the development of classical mathematics of probability and probabilistic reasoning to account for limitations of claims and partial ignorance in the classical information processing system over the epistemological space. But this development is restricted to the probability space without a trace of how we get there by simply assuming the existence of the possibility space. To apply the laws of thought and the mathematics of the classical system, an assumption of exactness is imposed on the information representation and exact stochastic conditionality is derived and measured in terms of exact mathematical probability. Neglected, then, is defective information structure due to all kinds of vagueness and vague symbolic representation.

Such a cognitive deformity in the knowledge-construction process is not limited to some knowledge sectors outside science. All knowledge sectors are characterized by this cognitive deformity and defective information structure as it has been argued in the discussions on the analytics in the theory of the knowledge square as a meta-theory on the theory of knowledge [R2.10]. It is the presence of the cognitive deformity and the defective information structure in the construction of the elements of the primary category of epistemic elements over the epistemological space that is the underlying force of Max Plank's statement: *But even the keenest logic and the most exact mathematical calculation cannot produce a single fruitful result in the absence of a premise of unerring accuracy. Nothing can be gained from nothing* [R8.53, p. 81]. The epistemic problem of exact science can be understood when one views the knowledge production as connected epistemic processes whose input at every stage is the information structure that initializes its search foundation, and where the knowledge system is self-correcting, self-refining toward exactness rather than some segment such as science being claimed as exact. Here, it may be pointed out that matter is energy, energy is information and information is matter in a continual process of quantity-quality transformation in a universal closure whose acquaintances are inexact to cognitive agents.

The information structure over the epistemological space is presented to cognitive agents as composed of objective and subjective information. The objective information comes to us as information signals from the elements of the universal object set which is then coded as subjective information through acquaintances that in turn is translated as empirical or axiomatic elements that enter the primary category of logical reality. The nature of empirical or axiomatic establishment of the primary category depends on the existing knowledge, social institutions of the knowledge-production process and within the ideological boundaries of societies. The ideological boundaries impose on the knowledge-production process presuppositions and subjectivity that are constructed from the socially held beliefs as to what is acceptable and what is not in terms of truth validations and researchable questions.

The belief in exact science in the knowledge-production process, without presuppositions or subjective characterizations, is a fiction of cognition and the knowledge-discovery process. In fact, the choice of the problem of interest and the needed methods and path of reasoning, conclusions to be drawn and interpretations of the results of the analysis and the synthesis into a body of an existing knowledge are subjective and ideologically directed. This is the decision-choice problem of all cognitive agents and in all sciences and knowledge areas. The basic idea is that exactness appears always in degrees and in every cognitive venture. It exists in duality with inexactness and connected by the logical continuum principle without which its meaning is opaque. The knowledge acquisition is always in relation to something. And this something must be identified, and where such identification is always decision-choice determined. The conditions of such a determination are themselves reflections of the cognitive activities on the basis of a subjective-objective phenomenon. This something carries the qualitative characteristics of vagueness and ambiguity of epistemic objects which have no unquestionable universal determination

The problem, here, is that the universe exists in continuity that obeys the continuum principle where the elements in it appear as relationally connected for continual substitution-transformation processes where nothing is lost except qualitative transformations in the ontological space. Each element has characteristics that allow a category to be formed as a primary category through cognition in relation to our linguistic structures at the level of possibility space into a possibility set. The category formation is not exact because of the universal relations, connectedness and continuity of their elements through their characteristics. As a category is formed in the epistemic space, every element cognitively manifests itself by some characteristics but not by all characteristics that establish the appearance of quality in a particular respect. The set of characteristics may vary from time to time for the same element and from a category to a category as seen and observed by cognitive agents. The problem is that any cognitive identity in terms of qualities is vague and hence each epistemic category formed manifests itself as a *fuzzy set* in our linguistic characterization where the defined boundaries are decision-choice determined to claim exactness. Such a decision-choice determination is *cognitive exactness* but not *ontological exactness* and hence has no universal claim. In fact, the epistemic exactness may substantially deviate from the ontological exactness. The difference is the *knowledge distance* due to ambiguities in thought in the activities involving vagueness and incompleteness in information structure.

Definition 1.1.2. Knowledge Distance

A knowledge distance is the difference between ontological exactness and epistemic exactness and measured by fuzzy-stochastic conditionality where such conditionality provides us with the risk of ignorance for the given phenomenon.

The closer the epistemic exactness is to the ontological exactness the shorter is the knowledge distance and the smaller is the human ignorance for any given phenomenon. Fuzziness, composed of vagueness and ambiguities, is an inherent property of knowing giving rise to inexactness in the knowledge systems, and hence any epistemic exactness is always decision-choice determined irrespective of the knowledge room under construction. This seems to be the universal principle of general cognition.

The fuzziness as an inherent property of thought is a strength of the property of relational motion of epistemic states and processes as well as in any language. In other words, the primary category of logical reality is also a fuzzy set where the only things that meet universality principle are existence, relations, continuity and substitution-transformation of elements. The driving conclusion in all knowledge production, including science that permits us with some level of comfort, in accordance with all foundations of thought, is that the placing of exact science in a prior logical foundation is simply a choice of convenience but can not be a universal claim. The possibility set with fuzzy categories must pass through the probability space in order to make its way to the space of the epistemic actual. Here, greater difficulties are encountered as thought is constrained by information limitation and vagueness in concepts, approximations of data and ambiguities in reasoning. Any attempt to classify science as exact is an attempt in placing limitations on the knowledge-production process and its enterprise, thus expanding the epistemic difficulties in human cognition in relation to the primary and derived categories of the knowledge-production process. Exactness in knowledge production is the ultimate goal to which cognitive agents work towards from ignorance in the domain of inexactness. By imposing conditions of exactness from the onset is equivalent in imposing the conditions of the ultimate and hence we strip the search process of its goal. The point here is the ontological information structure is exact. The epistemological information structure is inexact creating some epistemic difficulties for all areas of knowledge search and for all paradigms of thought in existence or constructible. Let us keep in mind that a paradigm is an epistemic machine for information processing to create a knowledge output. A paradigm, therefore, is simply an instrument to assist cognitive agents to process information into knowledge as well as to specify the conditions of knowledge acceptance.

It is these epistemic difficulties in establishing   the primary category in the epistemological space that may be seen in contemporary thinking as fuzzy set relative to the classical paradigm that propelled Max Plank to state:

> *If we seek a foundation for the edifice of exact science which is capable of withstanding every criticism, we must first of all tone down our demands considerably. We must not expect to succeed at a stroke, by one single lucky idea, in hitting on an axiom of universal validity, permit us to develop, with exact methods a complete scientific structure. We must be satisfied initially to discover some form of truth which no skepticism can attack. In other words, we must set our sight not on what we would like to know but first on what we do know with certainty.* [R8.53, p. 84]

From this impressive statement by one of the exceptional scientists and philosophers as well as one of the advocates of exact science, we are left to conclude that the differences between exact science and inexact science are matters of degrees in vagueness, ambiguity and subjectivity which are characteristics of qualitative dispositions for any quantitative disposition.

1.2 Epistemic Polarity, Duality and the Defective Information Structure

All the techniques and methods required for researches about knowledge on any phenomena are completely controlled by information incompleteness and vagueness that constitute defective information structure and give. This gives rise to the axiom of universal validity in cognition at the level of epistemology. The nature of knowledge production, in relation to exactness and inexactness, is defined by *three universal axioms of cognition*. These are: the axiom of *information vagueness*, the axiom of *information incompleteness*, and the axiom of *subjectivity and decision-choice rationality*. These axioms tend to characterize the primary category, methods of reasoning and culture of truth acceptance.

Our single task, but a difficult one, is a search for a framework or a paradigm of reasoning whose pillars are embraced with a logical strength that can deal with the qualitative characteristics of information incompleteness, vagueness in language, ambiguities in reason and subjectivity in judgment rather than assume them away in terms of a-prior exactness. This epistemic paradigm of knowledge production must have a complete system, with its logic and mathematics, which allows incompleteness, vagueness, ambiguity and subjectivity in interpretation and representation of information to be incorporated in the human cognition through the decision-choice actions of knowing. The knowledge production is a continuous process but not a discrete process. It is not exact and absolute but a self-correcting system with continual refinement toward perfection and exactness from the states of ignorance and inexactness.

It will be argued that the required paradigm in dealing with the conditions of the defective information structure is the fuzzy paradigm that must either replace or support the classical paradigm with the extreme cases for the knowledge production. It must then be related to rationality of knowledge production that must further be related to exact and inexact sciences for the unity of the knowledge production process, and then related to the fundamental discreteness and continuum in the epistemological space. It must be always kept in mind that exactness and inexactness constitute linguistic duality and reside in an epistemic unity, in that, every knowledge is simultaneously exact and inexact where the knowledge-production process is a refinement from the complete inexactness through a logical action by slowly stripping away the characteristics of inexactness and improving the degree of exactness as one journeys on the path of the knowledge discovery.

To explain the point of exact-inexact relationship, consider the concepts of polarity and duality within the theory of the knowledge square [R2.10]. There is the exact pole, let it be pole (1) and there is also inexact pole, let it be pole (2) where each pole contains a duality of exact characteristic set and inexact characteristic set at different proportions that provide the identity of each pole. Let $\mathbb{X}$ be the exact characteristic set and $\mathbb{X}'$ the inexact characteristic set. The sets $\mathbb{X}_1$ and $\mathbb{X}'_1$ are the exact and inexact characteristic sets contained in the exact pole respectively since $\#\mathbb{X}_1 > \#\mathbb{X}'_1$. Similarly, the sets $\mathbb{X}_2$ and $\mathbb{X}'_2$ are exact and inexact characteristic sets contained in the inexact pole respectively since $\#\mathbb{X}'_2 > \#\mathbb{X}_2$. In other words, the identity of each pole is defined by the relativity of exact-inexact characteristic sets. One cannot derive a quantitative relationship between $\mathbb{X}_1$ and $\mathbb{X}_2$ and between $\mathbb{X}'_1$ and $\mathbb{X}'_2$. Furthermore, let $\mathbb{U}_1$ and $\mathbb{U}_2$ be the total characteristic sets of the exact pole and inexact pole respectively; then $\mathbb{U}_1 = \left(\mathbb{X}_1 \cup \mathbb{X}'_1\right)$ and $\mathbb{U}_2 = \left(\mathbb{X}_2 \cup \mathbb{X}'_2\right)$. The relational nature of the exact-inexact polarity and duality are geometrically presented in Figure 1.2.1.

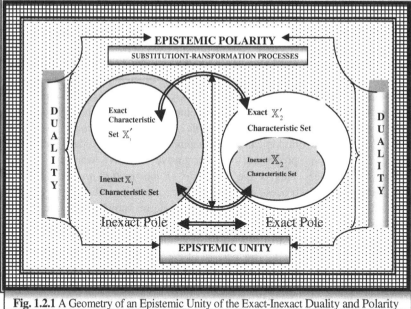

Fig. 1.2.1 A Geometry of an Epistemic Unity of the Exact-Inexact Duality and Polarity

The inexact pole, by linguistic characterization and decision-choice action, is that pole whose inexact characteristic set dominates the exact characteristic set in the pole while the exact pole is specified in likewise manner as that pole whose inexact characteristic set is dominated by the exact characteristic set. The point of implication is that any subject area of knowledge may be viewed as existing in exact-inexact polarity that contains exact-inexact duality under cognitive tension with logical substitution-transformation between inexactness and exactness in continuum. The defining condition of characterization is, simply, by degrees of exactness and inexactness contained in the subject area of science as determined by a decision-choice action with an appropriate qualification through the relative comparison of the set cardinality. Since each pole and hence each subject area of science or knowledge is under cognitive tension in logical substitution-transforming dynamics of inexactness-exactness duality, a subject area viewed as inexact today may one day be transformed into the class of exactness of science through decision-choice algorithm for sorting and filing depending on expansion of knowledge and technology.

The concept of exactness, as claimed by some scientists and philosophers of science and appropriated for some particular areas on the knowledge system, is derived from inexactness by subjective decision-choice action, and that, both exactness and inexactness exist in a dynamic unity of categorial conversions as activities in the epistemological space. There are parameters of acceptability of what constitutes exactness in any human cognition and in any given language. The foundation of our analytics and synthesis is based on four claims that constitute four principles of the knowledge-production process. The first claim is that the elements of nature are ontologically exact. This is the principle of *ontological exactness*. The second claim is that inexactness is epistemologically a cognitive phenomenon. This is the principle of *epistemological inexactness*. The ontological exactness and the epistemological inexactness are connected by defective information structure. This is the principle of *information connectivity*. The relationships between ontological and epistemological elements are established through information processing by cognitive agents working in the epistemological space. This is the *principle of information-processing* in the knowledge construction process with a particular epistemic toolbox.

Here, the attempt is to find a universal principle of knowledge production on the basis of methodological constructionism (forward reasoning) and methodological reductionism (backward reasoning). The claim here is that the universal principle of knowing irrespective of the knowledge area is composed of four sub-principles of ontological exactness, epistemological inexactness, information connectivity and cognition continuum. The exactness accorded to the results of the logical transformations of epistemological inexactness in the knowledge-production process is what we shall refer to as *epistemic exactness* which must always have *fuzzy-stochastic* or *stochastic-fuzzy conditionality*. Pure exactness is incompatible with the epistemological space and human thought and hence its assignment to a particular area of knowledge production is unjustifiable.

Every knowledge sector derives its conclusions and claims by a process from the epistemological space that is characterized by defective information structure. Any attempt to find a universal principle of a-priori nature for exactness and certainty is bound to be unsuccessful, and a search for such a universal principle will be nothing but a *phantom problem* in cognition. The knowledge-production process traverses through the epistemological space under the conditions of inexactness and information limitation where we can only derive exact and certainty equivalences.   Methods that define science and exact science may be claimed to be exact, and accepted only through a subjective action and a rationality of decision-choice process. Such a decision-choice process is driven by cognition and depends on an acceptable social framework of knowledge construction on the basis of information structure including transmission and reception. The new paradigm that we have referred to above must help to answer a number of questions in the enterprise of science and the knowledge-production process as they relate to quantity-quality relationship, vagueness, uncertainty, inexactness and subjectivity with neutrality of time. Some of these questions are stated in Table 1.2.1.

Table 1.2.1 Epistemic Questions on Quantity-quality Relationships in Science and Knowledge Productions

1. Are elements in the universal object set (the ontological elements) vague in their existence in which case the primary category of logical reality cannot be exact?

2. Can the elements in the universal object set whether vague or non-vague be known with certainty and exactness?

3. Are certainty and uncertainty characteristics of science and the knowledge production in general as seen from human cognition and the process of knowing or are they part of ontology?

4. Can we construct an axiomatic system about the elements in the universal object set that will have information belief such that vagueness, exactness and certainty are ontological rather than epistemological?

5. Do we have doubts of knowledge that is claimed to be of scientific truth and is this doubt due to information limitation or information vagueness or inexact reasoning or all of the above?

6 Can our measurements in knowledge take claim to exactness in units?

7. Can our current mathematical reasoning be modified to deal with both qualitative and subjective phenomena?

These questions go to the very foundation of exact science that Max Planck and others have sought to erect on the basis of a *principle of universal validity*. The general position of exact scientists is stated by Max Planck:

> *Now then, among all the facts that we do know and can report to each other, which is the one that is absolutely the most certain, the one that is not open even to the most minute doubt? This quest admits of but one answer: 'That which we experience with our own body'. And since exact science deals with the exploration of the outside world, we may immediately go to say: They are the impressions we receive in life from the outside world directly through our sense organ, the ears etc. If we see, hear or touch something, it is clearly a given fact which no skeptic can endanger.*
> [R14.87, p. 85].

This position of Max Plank is an affirmation of the empiricists' initialization of the primary category of reality through the sense data on the basis of which the derived categories can be abstracted by logical constructionism. This is another way of stating the position of Amo Afer who states that:

> *There is nothing in the intellect that has not already been in the senses. Now there is nothing in the senses, i.e. in the sensory organs, that has not already been in sensible things from which are distinguished things not perceptible to the senses. Nothing can be inferred therefrom except the operation of the mind, sensation and the thing itself. Now that a basis of enquiry has been thus established, learning is objectively either real, sensory i.e. historical and experimental, or intentional: it thus concerns respectively, (1)the intellect, (2) the will, (3) habitus and action, **i.e.** it is either logical, or moral or pragmatic* [R14.2, p102].

The statements by Amo and Planck give an assurance to an empirical foundation of science and the construction of the primary category of logical reality. Amo's statement adds to the assurance that the construction of the primary logical category from axiomatic foundation is also empirically derived but not necessarily experimentally derived. There is no assurance, however, that the elements of the sense data are exact. However, Max Planck points to cognitive position that: *The content of the sensory impression is the most suitable and unassailable foundation on which to build the structure of exact science.* The sensory impressions are characterized as *the sense world* which is claimed that: *The sense world is that which so to speak, furnishes science with the raw material for its labor* [R14.87, p.86]. The claim of building the foundation of exact science and any exact knowledge sector, and the success of this claim rest on the basic conditions that:

1) the primary category and its linguistic representation must be exact; 2) the possibility set from the possibility space must be exact; 3) the probability set in the probability space must be exact;4) the measurements in all these spaces must be exact; and 5) the methods of reasoning in abstracting the derived categories of knowledge must be exact. If the primary category and its linguistic representation (symbolism) are exact, the possibility set is exact, the probability set is exact, the measurements over all the relevant spaces are exact and the laws of thought are exact then the categorial conversion of logical categories are exact leading to exactness of derived category of knowledge. These conditions are difficult to meet by human artifice even if it is possible. They require that the epistemological space in which cognitive agents work contains non-defective information structure on the basis of which both exact symbolism and exact laws of thought are developed. This is not consistent with the world as we know. The exact symbolism may be related to the notion of the fundamental discreteness, digital processes and the classical laws of thought. The point of denial of the exact science does not mean that we cannot use methodological exactness and the tools of fundamental discreetness and digital processes for establishing scientific truth.

The task of the claimers of exact science, then, is to design a justified belief system in support of the exactness of the primary category of logical reality whose information is exact. It is here that Amo provides a justified belief for the empirical foundation of the primary category and Max Planck claims its exactness. What is presented to the sense organs is nothing more than a representation as models of what is represented, and if every representation is Russell-vague [R14.100], then all the models of sense organs are epistemologically vague and the primary category as an empirical foundation of science cannot take claim to exactness if a justification of its data input is sense-organ based. This is also affirmed by Max Black who states:

> ... all symbols whose application involves the recognition of sensible qualities are vague, and a typical case is constructed for convenience of reference. Vagueness is distinguished from generality and from ambiguity. The former is constituted by the application of a symbol to a multiplicity of objects in the field of reference, the latter by the association of finite number of alternative meanings having the same phonetic form; but it is characteristic of the vague symbol that there are no alternative symbols in the language, and its vagueness is a feature of the boundary of its extension, and is not constituted by the extension itself [R19.4 p.430].

Furthermore, the concepts of exactness and inexactness are qualitative variables or linguistic variables that take on linguistic numbers or quantities in representations that are determined by human decision-choice action in the knowledge-production system. The concepts in Max Planck's statement: *most certain,* and *the one that is not open even to the most minute doubt* are all vague and fuzzy variables that create more problems for the claim of exact science.

In other words, science and its knowledge-production process contain various degrees of inexactness and exactness where, among other things, the enterprise of scientific activities is an asymptotically continual transformation from epistemic inexactness to epistemic exactness which is then to be verified against the ontological exactness. But this holds for all areas of knowledge. The knowledge-production system is, therefore, a *fuzzy system that is self-organizing, self-correcting, self-improving* and characterized by defective information structure, and whose progress, mutation, differentiation combinatorial innovations and amalgamations require an alternative approach to the information-decision-interactive processes to the enterprise of the knowledge-production system. In this framework, science, the whole enterprise of knowledge production and their quality control come under the critical scrutiny of fuzzy paradigm composed of its logic, mathematics, laws of thought and its rationality where exactness is a derivative from inexactness by decision-choice actions to establish its linguistic quantities within acceptable boundaries for comparative analysis as well as to deal with *Zadeh vagueness* as the systems complexity increases [R19.77]. As it has been argued in [R2.10] and will be expanded in this monograph that there is an irreducible core of vagueness or inexactness in every knowledge area. As the system increases in its complexity, the irreducible core of inexactness increases and renders any method of exact representations of the system ineffective except by stripping the system of its essential properties that give it its qualitative disposition. The irreducible core of inexactness is related to the irreducible core of uncertainty which is related to the irreducible core of systemic risk in the knowledge-production enterprise.

The logical development of the claim about inexactness of science and that of the whole knowledge production and its enterprise as a fuzzy system is through the development of the theory of the knowledge square that is presented in [R2.10] where there is a cognitive interconnectedness among the universal object set (cognitive potential space), the possibility set (possibility space), the probability set (probable world) and the space of epistemic actual (space of cognitive reality). The connecting cords are information-decision-interactive processes working through a epistemic technology which is itself designed by cognitive agents. The vagueness of the representation, the ambiguities in the represented and the approximations in the reasoning in the process of obtaining the derived categories through the use of Aristotelian logic with its law of excluded middle, represent extreme cases of complete inexactness on one hand and complete exactness on the other. In this respect, the classical paradigm presents inexactness and exactness as dualism with excluded middle rather than duality with a continuum under cognitive tension with its internal conversion moment that admits of the emergence of exactness from inexactness through a logical perfection as an epistemic technological process, and where the permissible limits of variations of acceptability of degrees of exactness are imposed by a decision-choice rationality without allowance for subjective judgment.

Chapter 2
The Laws of Thought and Exact Science

Given that the justified belief in support of exactness of the primary category of logical reality is accepted, we still have a problem of establishing the exactness of logical transformations among epistemic categories as well as to establish the conversion chain of derived categories by the method of constructionism. Let us keep in mind that the acceptance of exactness of the primary category means that the defective information structure over the epistemic space is due to incompleteness and free from defectiveness due to vagueness and ambiguities. In this way the possibility space is assumed to be exact and so also the incomplete information in the probability space. The first step is to present an exact representation of the elements in the primary category that is empirically or axiomatically defined on which the laws of thought will be applied. The representation involves a language presentation of ideas and meanings through words or symbols for analysis and synthesis to arrive at the derived logical categories. Examples of the primary and the derived categories in mathematics are that the advanced mathematical levels are derivative from the basic number system which is a derivative from our linguistic number system.

2.1 The Primary Category and Information Representation

The initial concern for categorial conversions of logical categories is the information representation of elements in the primary category. The representation takes the form of information about concepts and ideas as established by words and symbols about the elements in the primary category. For matters of distinction, the objects in the primary category require the word-object definition and may be represented by symbols for defined operations in accord with conversions as established by the language of a particular knowledge construction or reduction. For the exact science, this representation is claimed to be exact and rooted in empirical or axiomatic foundation. However, for this representation to be exact, the word-object definition must be exact. Thus the construction of the primary category for the development of exact science requires that observations, experiments and sensory experiences must be exact and their definitions must be exact. The requirement is that the *definiendum* (that which is to be defined) must be exact through its method of cognitive existence; and the *definiens* (that which defines) must be exact in representation of the *definiendum*. The condition for the

K.K. Dompere: Fuzziness and Found. of Exact and Inexact Sci., STUDFUZZ 290, pp. 15–33.
springerlink.com   © Springer-Verlag Berlin Heidelberg 2013

achievement of the exactness in representation requires the equality between *definiendum* and *definiens* in which case explication is not needed.

In mathematical terms, this is a point-to-point mapping as it has been specified in the discussion in [R2.10]. The point-to-point mapping by a definitional function is only achieved in the world of conceptually neutral symbols that come under specified rules of operations in particular laws of thought and where a definitional mapping from a term to itself presents the term as self-defined in its communication and language. These hold to be valid for mathematical and logical symbols that are concept-free and hence pure mathematics, just like symbolic logic, provides us with no knowledge in and from the universal object set except as a means of studying laws of thought and corresponding computing systems. Thus mathematics becomes part of the class of theory of formal languages that is logically self-contained. In this way, mathematics, as an exact science, is only useful in teaching us the science of reasoning with symbols and prescribed rules of thought under strict conditions of exactness. Problems are, however, encountered when concepts are assigned to the symbols and the symbols are not concept neutral. As applied to the elements of the primary category of logical reality, the *definiens* is a linguistic thinking device that enables the user to distinguish, identify and construct the primary category. Here, aspects of word-object and real definitions such as lexical and stipulative may be discussed. The point under discussion can be carried on without them. For extensive discussions on the theory of definition and explication (see[R14.40][R14.93]). To unravel the difficulties involved in the relationships among the primary category, inexactness and the series of derived categories and the claim of exactness of science consider the conceptual, definitional and relational dynamics of *mater-energy-information* trinity whose relational geometry to the choice of the primary category   is illustrated in Figure 2.1.1.

A number of disturbing questions tend to arise in the study and understanding of ontological categorial conversions that will help us to initialize the epistemic journey in the epistemological space. Is matter energy or is energy matter? Is energy a derivative from matter or matter a derivative from energy? Is energy information or is information energy? Is energy a derivative from information or information a derivative from energy? Is matter information or is information matter? Is matter a derivative from information or is information a derivative from matter? Matter, energy, information and their relational structures are ontological aggregate elements that exist in exact and perfect states of being in the ontological space. How exact are these concepts in the epistemological space? How many definitions of energy do we have? How many definitions of matter do we have and how many definitions of information do we have? How are these concepts related to the conditions of epistemic understanding and the development of the knowledge house and the expansions of the rooms? How may these concepts be symbolically represented in relation to exactness and inexactness and what laws of thought must be used to operate on them? Do these elements present themselves as discrete (digital) or continuum (analog)? These questions acquire potency and cognitive relevance in the epistemological space where they point to possible relationships that will allow an epistemic mapping of the sets of epistemic elements into the ontological space.

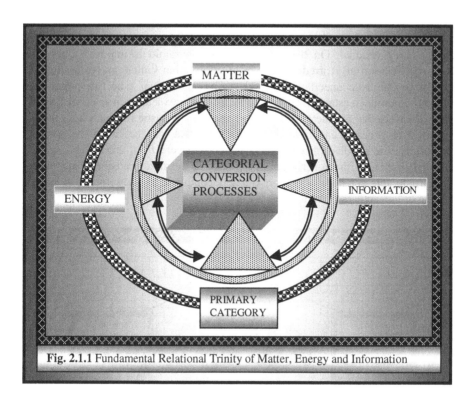

Fig. 2.1.1 Fundamental Relational Trinity of Matter, Energy and Information

It is very difficult if not impossible to guarantee the exactness of representations and definitions of elements represented in the primary category in the epistemological space since definitions and representations belong to linguistic categories that always carry with each definition a *residual vagueness* such that the *definiens* is not isomorphic representation of the *definiendum* in all knowledge constructions, except the set of the definiens includes the definiendum, in which case the definitional mapping meets the conditions of the Brouwer's fixed point theorem. This residual vagueness, we shall call *fuzzy residual* in definitions and representations. It is this fuzzy residual that specifies the boundaries of the power of human judgment in the applications of the laws of thought as well as ensure the beauty and usefulness of languages as reporting instruments of knowledge through communication. The fuzzy residual is the irreducible core of vagueness, ambiguity and approximation in the knowledge-production process and hence it is the *irreducible core of fuzzy uncertainty*. Corresponding to the residual vagueness is the residual information limitation that establishes the *stochastic residual* which is the irreducible core of residual information limitation that gives rise to *irreducible core of probabilistic uncertainty*. The cognitive geometry that relates these concepts to the understanding of the exact-inexact duality is presented in Figure 3.2.1 in a pyramidal logic where the immediate relational concepts are taken in three manifolds. The first pyramidal relation is a) irreducible core of vagueness, ambiguity and approximation, b) irreducible core of fuzzy uncertainty and c) fuzzy residual. On the top of this first pyramid is imposed a second pyramidal relation of a) irreducible core of information limitation, b) irreducible core of probabilistic uncertainty and c) stochastic residual. The two pyramidal relations combine to establish an irreducible core of inexactness in sciences and all areas of knowledge production. The fuzzy and the stochastic residuals are always combined to generate an irreducible inexactness that may be called the core of fuzzy-stochastic residual of inexactness which will vary from knowledge area to knowledge area. With properly defined functional forms of categorial mappings these irreducible cores may be approximately computed to allow for the possible ranking of inexactness of knowledge areas.

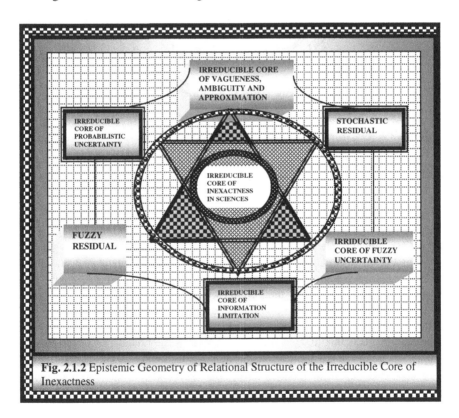

Fig. 2.1.2 Epistemic Geometry of Relational Structure of the Irreducible Core of Inexactness

For exact science, it is not sufficient for the elements in the primary category to meet conditions of exactness of observation, experimentation and representation but must be measurable. This measurability requirement is stated by Max Planck as:

> *Since exact science deals with measurable magnitudes, it is concerned primarily with those sensory impressions which admit of quantitative data--- in other words, the world of sight, the world of hearing and the world of touch. These fields supply science with its raw material for study and research, and science goes to work on it with the tools of logically, mathematically and philosophically reasoning.* [R14.87, pp. 87-88].

The introduction of measurable magnitudes brings into the knowledge- production process a difficulty of accuracy of our measurements in the exact science and in the general knowledge production. There are many concerns of accuracy in measuring sense data and the units of measurements used that have been discussed in the previous [R2.10]. At quantitative levels, these measurements are approximations that are taken to be exact by decision-choice actions. The requirement of measurable magnitudes, further, complicates the understanding of the full scope of knowledge production process by mainly concentrating our attention on *time and quantity*, and to the neglect of the fundamental order of *quality and time* in categorial conversions, for example chemical interactions, energy transformations, energy-informational transformation and social change or color changes through mixing. The Knowledge-production process that allows us to reduce ignorance about nature and society requires us to deal with *quantity, quality and time* where we must study both the quantitative and qualitative motions that define the trajectory of complete changes and transformations as time progresses. This implies that the task of the knowledge production is the study not only of states but also of processes in time-quantity and time-quality spaces as well as quality-quantity space. At the level of measurable quantities, the essential point that must be taken note of is the notion that our technology of measurements is interactive with the progress of human know-how in such a way that what is not directly measurable today may acquire measurability and quantitative characteristics as knowledge accumulates. Importantly, the relational structure of matter-energy-information trinity and the conditions of convertibility are established through qualitative motions that carry with them conditions of inexactness especially when these basic epistemic items are related to power-force-work trinity.

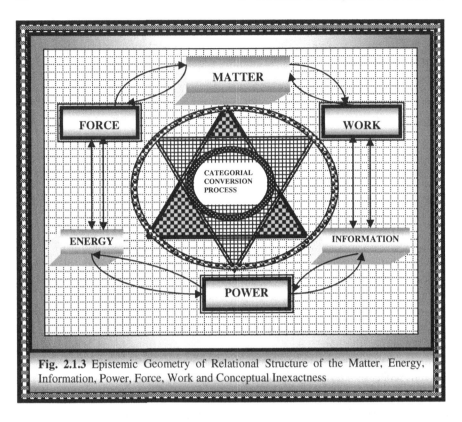

Fig. 2.1.3 Epistemic Geometry of Relational Structure of the Matter, Energy, Information, Power, Force, Work and Conceptual Inexactness

The claim of the possibility of measurability of currently immeasurable qualities and linguistic quantities is rooted in the concept of natural duality where quantity and quality exist in unity under transformational tension in a continuum, and where quantity has a qualitative disposition and vice versa of elements in the universal object set. Every epistemic element in the knowledge-production process exists in a quality-quantity duality. What is claimed to be inexact science because the elements in the primary category of logical reality fails, by human decision-choice action, to meet the essential characteristics of measurability and other supporting attributes may come to acquire the measurability properties as the knowledge structure expands and our technological know-how conquers new frontiers of measurability and quantification. The lack of measurability in any subject area is not a deficiency of inexactness of the knowledge area but a failure of ingenuity of human logical action in the area due to increasing complexity in quality-quantity conversions. At this point, it is useful to point out that the introduction of quality introduces added vagueness, ambiguity and subjectivity that amplify the fuzzy residual in measurements and representation. For example, the measurability attributes of energy, matter, information, work, power and force are human decision-choice determined and are no different from taste, beauty and others as qualitative elements, except we have found a way to assign quantity to their qualitative attributes.

The basic characteristic of all subject areas of knowledge that take claim to exact science is that they have done away with quality or qualitative attributes of representation in these areas of the knowledge-production processes and where the fuzzy residuals in their measurements and representations are assumed away or taken to be negligible or unimportant, and hence an exactness is claimed. In this framework, a linguistic framing in a conceptual system of meanings of questions and answers within contexts are stripped off their importance. Vagueness is an essential and important qualitative characteristic of all linguistic systems. Ambiguity is also an essential and indispensible qualitative characteristic of communications among cognitive agents. Approximation is an importantly inherent attribute of cognitive calculus in the human information processing activities over the epistemological space. And judgment is also an inherently distinct feature of all decision-choice actions of cognitive agents to arrive at claims. The presence of vagueness and limitations in the information structure, ambiguity in language and reasoning, approximation in measurement and cognitive calculus and subjectivity in human decision-choice actions must not be viewed as deficient but rather important components in the information-knowledge process. The presence of these qualitative characteristics and their increasing intensity in different areas of knowledge production must be seen as a problem that challenges the cognitive capacity of knowing agents. The solution to the problem of inexactness, therefore, is not how to assume away these qualitative attributes of the thought formation but rather how to provide an adequate symbolic representation and laws of thought that will accommodate vagueness, ambiguity, approximation and subjective phenomena. In these connections see the discussions in [R14.14] [R14.15] [R19] [R19.4] [R19.22] [R19.25] [R19.37] [R19.44] [R19.45] [R19.58].

2.2 The Exactness of the Classical Mathematics and Its Applications in Exact Science and Knowledge Sectors

The developments of classical mathematics and symbolic logic are based on representations where the definitional function is a point-to-point mapping, or a fixed point mapping in that the set of the definiens contains the definiendum and hence x is x or $x = f(x)$ where the x functions as both the definiendum and definiens to establish an identity. In other words, x is, by definition, mapped onto itself and hence needs no explication. The symbols are exact with exact variables devoid of subjectivity and conceptual context. The representation in exact mathematical development, its category formations including sets and groups, its axioms and its theorems are only representations of quantity and time in an exact form which presents to us the quantity-time phenomenon including space-time relations and where every element in a set has the same fixed quality without shade.

This kind of exact representational system, we have argued, requires an important assumption where the information structure in the epistemological space is taken to be non-defective or perfect in the sense of being exact and complete. When the representation is extended into conditions of information incompleteness the assumption of exactness is retained. Given the assumption of the perfect information structure, the only thing needed is to design rules of thought that will provide exact operations on these exact representations where each symbol is conceptually neutral. Mathematics in this way reveals no knowledge about nature and society; it mainly provides us with exercises in thinking within a specified paradigm. Since its representation is devoid of subjective phenomenon and conceptually neutral, its rules of thought are only applicable to quantitative but not qualitative dispositions. It is on this basis that mathematics is claimed to be an exact science and applicable in all language systems and all knowledge areas whose fundamental assumptions can be made parallel to those of mathematics. In this connection, the observation by Max Black is instructive.

> *For mathematics is the study of all structures whose form can be expressed in symbols, it is the grammar of all symbolic systems and, as such, its methods are peculiarly appropriate to the investigation of its own internal structure. But the structure of mathematics, though implicit in its theorem, is not clearly shown and tends to be confused even by those who are most familiar* [R14.6, p.4].

The problem here is that mathematical knowledge is knowledge of its internal structure as a system of exact symbols, axioms, definition, operators, rules of theorem constructions, proofs and derivations. Here, disagreements arise among the three types of views on mathematics as seen by the positions of Logistics, Formalists and Intuitionists. At the level of *logistic view,* mathematics and logic are indistinguishable [R14.6] [R14.][R14.R8][R14.97][R14.99]. At the level of *formalist view,* mathematics is a science of structural properties of objects where the real number system represents the simplest and hence is concerned with characteristics of symbols completely independent of their meanings and concepts [R14.6] [R14.40] [R14.75] [R14.76] . At the level of *intuitionist view,* mathematics is seen as part of thought and hence a language of thought. It is, thus, not exact with timeless truths and hence contains vagueness and ambiguities [R14.14] [R14.15] [R14.30] [R14.53]. The position of the intuitionists is expressed by Max Black as:

> *Mathematics, regarded as a body of knowledge, grows, it is a becoming, a process, which can never be completely symbolized___ and even this manner of regarding it is perhaps dangerously abstract. Mathematics should be regarded as a social activity by which individuals organize phenomena in their most general aspect to satisfy their needs. Hence it is not enough to have a symbolism for mathematical thoughts; they are independent of the particular language used to express them. What is absolutely necessary is that the language should significantly express thoughts* [R14.6, p. 10].

The claim of mathematics as an exact science seems to flow from the conceptual position of the formalist where mathematics as a science is devoted to the study of properties of symbols without concepts. In this formalist system, therefore, problems of meaning are encountered when these symbols, axioms and theorems are attached with concepts and meanings from which thoughts are to be derived, analyzed and assigned conceptual understanding in the epistemic reality. It is here that application of classical mathematics in different areas of sciences cannot claim exactness. The Classical mathematics, as viewed from the position of the formalists, provides us with rules of organizing thought in its own internal system. It has no representation in reality that relates to sense data as suggested by Amo [R14.2] and Planck [R14.87]. Its truth can only be shown and agreed upon in its own internal system with its rules of reason, and since it is content free, it devoices itself from the principles of framing where meanings are contextual that give rise to ambiguity and vagueness at the presence of quality. In this way, the classical mathematics deals with the defective information structure by assuming away the linguistic inexactness of thought by imposing perfect information condition where we search for exact symbolism and rigid system of thought. In this way exactness is taken as a-priori condition of science.

On the other hand, if mathematics, as a system of reasoning and organization of ideas, is viewed from the intuitionist position where mathematics is seen as part of the development of human thought as well as part of general language of thought, then mathematics presents us with a broad general linguistic and epistemic space of knowing where, quality, vagueness and ambiguity are seen as essential characteristics of the information-knowledge process, and where judgment presents itself as a subjective phenomenon that relates to the decision-choice processes in cognition. The intuitionist constructive position presents us with a broader conceptual framework of mathematics whose structure can assist in creating rigor in thought for applications in different areas of the knowledge-production process through appropriate symbolic language of information representation at the presence of *defective information* structure, where exactness is arrived as part of the process of thought by decision-choice rationality. It helps us to advance laws of thought beyond sense data and simple experiences as imperfect epistemic elements abstracted from perfect ontological elements by acquaintance.

Here, the study of mathematical and symbolic true-false relationships is seen as decision-choice determined as we journey through the penumbral regions of human-nonhuman interactive processes that present complexity and continuum in the *quality-quantity-time space* with cognitive tensions. Here, exact science is not claimed a-priori; exact science is a state toward which cognitive agents approach through continual refinements where the states of exact and inexact sciences are essential parts of the cognitive system. This intuitionist position has given rise to rigorous developments of *fuzzy paradigm* with its logic and mathematics. Here, we search for symbolic representation of defective information structure to initialize the epistemic work and develop rules of thought for the manipulations of vague symbols that represent concepts rather than exact symbols that are devoid of concepts. In this way, inexactness is a-prior and exactness is reached a-posterior [R14.14], [R14.15], [R14.31], [R19.4], [R19.7], [R19.33], [R19.58] [719.65]. In this setup, the analytical processing of symbols is also analytical processing of information to arrive at significant conclusions.

In the intuitionist conceptual framework, mathematics acquires powerful relations with all areas of knowledge production as well as provides them with some strength of analytical rigor and comfort that allow us to deal with matters in both quantity and non-quantity spaces of the elements in the universal object set. Furthermore, it allows us a breathing room to deal with the understanding of qualitative transformations in addition to quantitative changes in the explanatory sciences where improvement in *what there is*, is not contemplated and in prescriptive sciences where improvement of *what there is*, is contemplated. The intuitionist conceptual framework requires us to develop mathematical and logical symbols that can incorporate vagueness and ambiguity that capture quality and quantity with neutrality of time without externalizing the knowledge seekers from the information-knowledge production process by simply following the rules of exact rigid determination of the classical paradigm. The search for vague symbols is not that ontological elements are vague but rather the epistemological elements are vague by the nature of limitations of human language, acquaintances, observations and cognitive capacity as links are established between ontological and epistemic elements in cognition.

Given the defective information structure for processing and knowledge production in the epistemological space, two interrelated problems are to be resolved in cognitive systems. The defective information structure contains qualitative and quantitative characteristics as time progresses. One of the problems is to find symbolic representations that will allow vagueness and ambiguity (inexactness) of the elements in the information structure to be incorporated as essential attributes in the symbolism. This is the problem of information representation. The second problem involves the development of a paradigm of thought with its logical and mathematical system that will allow the manipulations or the processing of these *vague symbols* to derive reasonably crisp (exact) conclusions with permissible boundaries of thought that relate to symbols which are not content-neutral but content-dependent. This is the problem of paradigm development. The appreciation of the information-representation problem and paradigm-development problem in relation to the presence of a defective information structure for cognition over the epistemological space have been reflected on by different thinkers in different ways and over different knowledge areas.

>*One of the most fundamental facts of experimental physics a student has to learn at the beginning of his studies is that no measurement is precise. He learns to estimate the margin of error of the measurements he carried out. What about these imprecisions and so-called errors of measurements*
>
>*A theoretical physicist will realize that there is another field of imprecision. He will realize that every mathematical theory used as a picture of reality cannot be regarded as a precise picture. He will realize that there is no difference in principle*

> *between a so-called exact theory and an approximation picture of reality. What is the meaning of approximation and imprecision in this field?* [R19.37, p. 7].

When we leave the realms of concept neutrality in symbolic thought and hence concept-neutral logic and mathematical symbolism and enter the concept-dependent areas of thought which applies to all areas of knowledge production, except the formalist mathematics, we can only think of vague but not exact symbolism whose manipulations are inexact and error-dependent with ambiguities that call for incorporation of human judgment into the laws of thought. The classical paradigm with its laws of thought and mathematics are only helpful at the true-false extremes but not reasoning between the true-false extremes. Here, it must be emphasized, as it has been argued in [R2.9][R11.21][R11.22] that the problem is not errors in logics or implementations in logical rules that may give rise to paradoxes. The problem is an epistemic one that challenges cognition to find a symbolic system that can incorporate vagueness and ambiguity in concepts and meaning contained in linguistic variables as seen in the fuzzy or inexact spaces as we seek knowledge relationship between the ontological and epistemological spaces. Such a symbolic representation that incorporates vagueness and ambiguity is the *fuzzy symbolism* capturing the defective information structure in the epistemological space as an alternative to the *classical exact symbolism* operating on exact information structure, by creating exact epistemic elements with exact mathematical and logical spaces, where the exact epistemic elements are manipulated with exact laws of thought to derive exact rigid propositions over the epistemological space.

The fuzzy (vague) symbolism is used to create inexact epistemic elements by developing inexact mathematical and logical spaces as a natural extension of the defective information structure that is available for processing. Given the fuzzy symbolic system, the requirement is to develop laws of thought that are consistent with the new logical and mathematical spaces that allow reasoning to map inexact space to conditional exact space for all knowledge sectors. This conditionality, we have named as *fuzzy conditionality* that allows us to derive *exact-value equivalences* from inexact space through membership characteristic function and fuzzy-decomposition process. The problem confronting cognition boils down to constructing a universal system of representation, logic and mathematics that can incorporate inexact and exact quantitative and qualitative structures for reasoning where decision-choice actions of cognitive agents with some subjective judgments are incorporated into the knowledge-production process in the fuzzy space of thought.

The fuzzy space of knowledge production presents some epistemic difficulties for the use of the classical paradigm of reason which is developed to be applicable to an ideal universe where there are either no qualitative characteristics or if there are, they are assumed to be constant to allow stable laws of thought to be abstracted. Here, the following statement is useful for our understanding as well as to note.

> *The notion of an ideal universe in which the laws of logic and mathematics have unconditional validity having been rejected, it remains to show how the undoubted usefulness of the formal sciences in a field of vague symbols can be explained by an extension of the method already sketched in the earlier sections* [R19.4, p.451].

Interestingly, Max Black comes to recognize that the classical dualism with mutually exclusive opposites must be abandoned where he states:

> *The generalization of usual notions of implication or negation of propositional functions of a single variable will be relations connecting the corresponding values of the consistency arrangements in two propositional functions. It follows from the definition of the consistency function that if* $L(x,C)$ *and* $L(x,C')$ *for the same* x*, the product of, the two consistencies,* C *and* C' *is in unity. Thus the principle of excluded middle is replaced by the operation which permits the transformation of* $L(x,C)$ *into* $L\left(x,\dfrac{1}{C}\right)$ [R19.4, p.452].

Max Black, then, introduces the concept of unity of the opposites in duality and suggests a principle of continuum that permits transformations between the logical poles. It is useful to point out again from the analysis in [R2.10] that the concepts of duality and continuum project the existence of a *set* with infinite elements that is bounded within the opposites with implied conditions of belonging. The problem, in its essential form, is not different from the Cantor's continuum problem [R14.39]. The problem, then, is how to specify the *set* and the *conditions of belonging*. To resolve this problem, Max Black exists from the rigid constrained boundaries of dualism with mutually exclusive and collectively exhaustive negative and positive sets of characteristics. He then enters into the zone of duality with non-mutually exclusive but collectively exhaustive negative and positive sets of characteristics. It may be noted that the transformations between poles are infinite that requires epistemic interventions through decision-choice actions by the cognitive agents on the principle of continuum. This analytical approach internalizes the knowledge seekers as part of the knowledge-production process.

    In this way, the outputs of the world of knowledge enterprise, just like any of the human productive activity, become dependent on the thinking individuals obeying the laws of inexactness of thought whose objects are directly or indirectly related to the ontological elements, and who then determine what is cognitively acceptable in the space of vagueness, ambiguity and imprecision in order to claim exactness with some conditionality. In this respect, Max Black states:

> *For while the vagueness not only in terms of the premises but in their relations prevents us from asserting the conclusion of an argument in applied logic or applied mathematics without qualification as*

to the degree of consistency (whose amount depends on the precision of the terms and logical relations) the form of the transformation is independent of the actual consistencies provided we are satisfied with a final precision with which increases indefinitely when the precision of the premises increases. [R19.4, p. 455].

As we have argued, the claim of exactness requires a fuzzy conditionality as a qualification for its acceptance. The fuzzy conditionality depends on the degree of exactness assigned to the claim and it is obtained through the combinations of fuzzification-defuzzification process, decision-choice rationality and the principle of fuzzy decompositions that establish the conditions of acceptable degrees of exactness and inexactness in the epistemic element of cognitive interest. The fuzzy conditionality takes care of the Max Black's qualification as to the *degree of consistency* in precision and partition of exactness from inexactness [R19.4] [R14.6].

Speaking of imprecision between theory and reality and how to represent it in physical theories G. Ludwig states:

The infinite extension of X (space) is only a substitute for our ignorance regarding the extent of real space.

The fact that we substitute our ignorance by idealized structures in the mathematical theory shows that we do not regard the mathematical theory as a precise picture of reality. There is some imprecision in the relation between a mathematical theory and the reality we want to describe with the theory.

If there is no mathematical theory which is a precise picture of reality, we have to seek a mathematical structure suitable to describe the imprecision in the relation between the mathematical theory and reality. But since we do not know the magnitude of imprecision we prefer to formulate the "structure of imprecision" in such a way that the "imprecision" in the mathematical picture can be improved more and more without any finite limit. This last aim is again an idealization, necessary because of our ignorance concerning the finite limit of precision in the relation between mathematical picture and reality. In this sense we try to introduce an idealized "structure of imprecision" which compensates – or better – partially compensate the idealizations introduced before. What could be a suitable structure of imprecision in the space X, which partially compensates the idealization e.g. of the continuum structure of X?

It is my opinion that in mathematics is called a <u>uniformity</u> or uniform structure is suitable to describe the imprecision in the interpretation of a physical theory [R19.37, pp.7-9].

To resolve the problem of the imprecision, Lugwig introduces the concept of *imprecision set.* Here, the consistency principle introduced by Max Black takes the form of *uniformity principle* in the Lugwig's analysis. The implied set in Max Black is replaced by imprecision sets by Lugwig. The Lugwig's statement that, *the "imprecision" in the mathematical picture can be improved more and more without any finite limit,* points to a limiting process of *fuzzy tuning* with the presence of an irreducible imprecision that we have called the *fuzzy residual* in

Table 2.2.2 The Essential Characteristics of the Classical and Fuzzy Paradigms for Processing Defective Information Structure over the Epistemological Space

Table 2.2.2 The Essential Characteristics of the Classical and Fuzzy Paradigms for Processing Defective Information Structure over the Epistemological Space

THE CLASSICAL PARADIGM	THE FUZZY PARADIGM
1. Reject the defective information structure and create a perfect information structure by assumption of exactness and not necessary fullness to create an exact epistemological space for thought processing; 2. Develop exact logical and mathematical spaces from the exact epistemological space; 3. Create exact symbolism for exact epistemic elements and the primary category; 4. Impose the principle of excluded middle in reasoning and reject the simultaneity of true-false existence; 5. Develop classical exact laws of thought that accept a-priori exactness; 6. Use these exact laws of thought to develop an epistemic processing machine for the input-transformation and; 7. Create knowledge and claim it to be exact	1. Acceptance of defective information structure as essential characteristic of sense data and hence accept inexact epistemological space for thought processing; 2. Develop inexact logical and mathematical spaces from the inexact epistemological space; 3. Create inexact (vague) symbolism for vague and inexact epistemic elements and category; 4. Reject the principle of excluded middle in reasoning and accept the principle of logical continuum where true or false can simultaneously exist; 5. Develop fuzzy laws of thought that accept a-priori fuzziness and a-posterior exactness 6. Use these fuzzy laws of thought to develop an epistemic processing machine for the input-transformation and; 7. Create knowledge and claim it to be a-posterior exact with fuzzy conditionality under fuzzy tuning.

fuzzy logical system. The elements in the set of imprecise characteristics are reduced to increase the elements in the set of precise characteristics of an epistemic object through a process. It is interesting that Gödel's discussion on Cantor's continuum problem passes through the concept of set [R14.39]. The Max Black's *consistency principle* and Lugwig's *uniformity principle* as qualifications of the derived exact-value equivalence are replaced by the principle of *fuzzy conditionality* in the fuzzy paradigm of thought. The Black's set and Lugwig's set are replaced by a generalized set of *Fuzzy set* that must be related to the derived categories on the path to epistemic reality. We mat tabulate the similarity and differences between the classical and fuzzy paradigms in working with the defective information structure.

2.3   The Derived Categories and Exact Logic

Given the exactness of elements in the primary category of reality in relation to definitions and measurements, the next step in the construct of exact science is the construction of derived categories by a method of reasoning. The reasoning must be exact if the categorial derivatives are to carry the defined properties of exactness of the primary category of reality. Here, an appeal is made to the foundations of the classical paradigm composed of its exact logic, representation and mathematics. The question that arises is how is the classical paradigm related to reasoning in science where we must deal with concepts, contents, quality and meanings? To answer this question, let us look at what logic connotes.

Thoughts in exact sciences, like any knowledge sector, must satisfy reflections on logical reality irrespective of the language used. The construction of derived categories gives rise to the need of laws of thought or what is generally referred to as principles of logic or rules of reasoning. The principles of logic or reasoning involve two important steps in moving from the primary category of knowing to the derived categories of knowing. The first step may be viewed as a principle of constructing significant propositions that link elements in the primary category to the needed derived category or categories. The second step involves the principle of linkage determinations where propositions about elements in the derived category are examined to be consistent with, and flow from the propositions about the elements in the primary category. The two together constitute the principle of parallelism in epistemic construct that have been discussed in [R2.10] under the theory of the knowledge square.

In this respect, principles of logic and laws of reasoning consist of cognitive task of providing order and regularity into the elements in the primary category that either contains heterogeneous empirical elements from the sense world or axiomatic conceptual initialization that allows categorial conversion of the primary category into the derived ones. The principle of logic tells us nothing about truth-value of statements that are contained in the derived categories, except when they are verified against the elements in the primary category that is rooted in the possibility space. It may be noted that the principles of logic are merely laws of reasoning in the language of its residence. They are not reality; they are designed by cognitive agents to assist in information processing and agreements in thought over the epistemological space. They merely reflect the organization of the epistemic elements in the primary categories as they relate to the derived category of logical reality. They help to point to a picture of epistemic reality in thought in the derived category that is to enter the probability space and as seen from the elements in the primary category from the possibility space. Any principle of logic is a set of rules of reflection on the ontological reality in thought to the extent to which the elements in the primary category are reasonable representation of reality as viewed from the ontological space. Logic shows how to combine concepts in linguistic frames to construct thoughts and derive meanings under conditions of relationality in a given language that is relationally connected to the epistemological space. It is part of the epistemic input-output processing machine that converts information into knowledge.

The laws of thought are the same as they are directed toward creating models of reality as objectively reflected in the primary category. They are reflected in every thought process and expression of the language of their existence. The laws of scientific thought are, however, systematized, but not different from ordinary day-to-day thinking in a given language. They differ merely in degrees of refinement and accuracy that is accepted by decision-choice action of cognitive agents. The laws of thought in an ordinary language, just like in the scientific language, must help to process words and symbols in order to create thought and meaning. This is an affirmation that there is only one kind of a universal principle for categorial conversion from the primary category to the derived. The degrees of refinement and accuracy are reflections to overcome the inherent nature of vagueness in representations, inexactness in logical transformations, ambiguities in the language of thought and approximations in conclusions. The nature of the constructed laws of thought depends on the assumed information structure of the primary category on which the laws of thought are to be applied.

As the laws of thought relate to principles of cognition and information-processing capacity of cognitive agents, the attempts in specifying the requirements of exact science are reflected in the use of facts of experience to initialize the primary category, as a foundation for a unified comprehension of the world in which we perceive and seek an understanding. The inputs of the primary category are information summaries of cognitive elements for logical processing into derived categories. Since the epistemological space contains defective information structure to begin with, the epistemic elements in the primary category are vague and inexact which constrain precision while our limited information constrains certainty. In a search for exactness and certainty, exact science takes claim to empirical base of the primary category and the exactness of laws of thought from the classical paradigm with its logic and mathematics. The question of exactness and certainty of science *boils down to the question of whether there exist any knowledge element that is of ultimate certainty and clarity that defiles all doubts given human cognition, information-process capacity, language, technology and culture of knowing ?* [R19.58] [R14.97], [R14.100].

We shall refer to this as the *ultimate certainty-precise question* in the knowledge production. The ultimate certainty relates to information completeness and objectivity, while ultimate clarity relates to absence of fuzziness, composed of vagueness, ambiguities and subjectivity. The answer to this question is fundamental to the claim of exact science and the understanding of the conditions that give rise to the development of fuzzy paradigm with its logic and mathematics as well as the understanding of the intuitionist critique of the formalist school of logic and mathematics. The fuzzy and intuitionist paradigms suggest that ultimate certainty-clarity knowledge exists as a state that knowledge seekers work towards through a process. The state of the ultimate clarity is attained through a *fuzzy tuning* and a limiting process to obtain exact-value equivalence; while the state of ultimate certainty is attained through *probabilistic tuning* and a limiting process to obtain certainty-value equivalence. The lack of their attainment is explainable by the presence of the irreducible core of *epistemic uncertainty* composed of the composite sum of irreducible cores of *fuzzy residual* and *stochastic residual* of uncertainties that give rise to risk in knowledge production and decision-choice systems.

Max Planck takes the position that exact science as a knowledge sector is constructible on the basis of information from sense organ [R14.87]. Bertrand Russell doubts that this exact knowledge is constructible [R14.100]. Amo affirms sense data as basis of knowledge construction but does not claim exactness [R14.2]. Max Planck seems indecisive, in this respect, in that he accepts the presence of vagueness in the knowledge-construction process and some level of certainty in mathematics whose very nature may be viewed as a grammar of all symbolic systems as seen by Max Black who states:

> For, when faced with the difficulty of clarifying existing knowledge, the temptation is greater to find compensation in admiring the complex structure which represents partial success and supplement it by unwarranted extrapolation [R14.6, p. 5].

All these call for some compensatory process in which Bertrand Russell calls into play the Axiom of Reducibility and Berkeley call into play the divine intervention [R14.6]. In terms of the theory of knowledge square that we have discussed in [R2.10] the Axiom of Reducibility may be viewed in terms of developing an *Axiom of Reducibility of fuzzy residual* in the process of claiming exactness and *axiom of reducibility of stochastic residual* also in the process of claiming certainty in the knowledge-production process.

There are many reflections that can be brought against the positive and negative claims about the pillars and foundations of exact science to maintain the notion that its foundation is strongly embraced. The primary category, constructed on the basis of sense data that reflects experience, does not have an unquestionable final character or universal principle. The sense data is a product of a process that connects cognitive agents to the elements that are sensed in the universal object set from acquaintance and awareness. These acquaintances and awareness become codes in the sense data as experience and subjective information on the basis of which thoughts are formed and projected as propositions whose truths or realities are to be subjected to tests in the probability space for acceptance in the space of the cognitive actual and an epistemic reality. Experience or sense data evolves as a process of cognitive relation that is subjectively reflected as codes of systems of sensation, memory, imagination, beliefs and others in the universal object set or reflections about some ontological elements. This process is buried in information-knowledge duality with logical conflicts that are produced and settled by cognitive activities through substitution-transformation processes of logical categories in the space of true-false duality. The information-knowledge duality is past-present-future (*Sankofa*) connected [R2.9] [R.11.22] [R14.29] [R14.100, p.12] in that the elements in the universal object set exist in interconnections and in substitution-transformation states, where manifestations of their attributes may be observationally unique to different cognitive entities as they apply the laws of thought in moving from information to knowledge claims under the *fuzzy continuum principle* in the epistemological space. This continuum principle is related to the fundamental discreteness, digital and analog representation of information and the laws of thought.

The laws of thought provide rules of logical dynamics that connect the epistemic elements in the primary category to the epistemic elements in the derived categories by the methodological constructionism or connect the epistemic elements in the derived categories to the epistemic elements in the primary category by the methodological reductionism. These laws of thought or laws of mathematics are not reality but laws of designing models and representation of epistemic reality in thought. The linguistic representation of the primary category and the cognitive subjectivity of reasoning are such that these laws cannot be claimed to be exact neither the results can be claimed to be absolute in truth. Every element of human cognition must then be seen in terms of *partial exactness* and *partial truthfulness* and hence the exact logic of science is a claim by decision-choice action in the spectrum of inexactness to exactness as has been explained in [R11.22] [R19.4]. The statement by E. Grafe is useful.

> From an epistemological view, classifying a statement as vague means to judge the statement in question to be a mixture from partial knowledge and partial ignorance. Accordingly it seems desirable to describe the boundary between knowledge and ignorance hidden in the vague statement [R19.22, p.113].

Let us examine the role of logic as a connector and an instrument for the production of categorial moment in creating logical transformation-substitution processes between the primary category and the set of categorial derivatives and the claim of exact science. There are a number of epistemic problems about the claim of exactness of a particular branch of the knowledge-production process when the premise of ontological exactness in the universal object set is accepted for the epistemic construct of the primary category in the possibility space. In [R2.9] and [R2.10] the concept of the knowledge square is introduced and discussed to provide an understanding that all knowledge about reality is produced by cognitive transformations from category of false or ignorance to category of truth or knowledge and vice versa, and that this categorial conversion takes place through an appropriate reasoning or fuzzy laws of thought in continuum where complete certainty and exactness are perfect states toward which our search for knowledge ends. These perfect states are unattainable. Their conceptual existence provides the incentive for the continual knowledge search by improving the fuzzy and stochastic tuning in attempts to reduce the knowledge distance. In other words, any state of knowing is simply a mixture of partial knowledge (exactness) and partial ignorance (inexactness). The nature of the approaches of the classical and fuzzy paradigms in dealing with the defective information structure and characterization of the epistemological space is such that the logical space in which fuzzy paradigm works is larger than that of the classical, leading to the condition that certain classical mathematical and logical propositions including the claim of exactness and exactness of science do not hold in the space of fuzzy paradigm. Let us turn our attention to the relationships among primary and derived categories and scientific world pictures.

Chapter 3
The Primary Category, Derived Categories and Scientific World Pictures

Any cognition in the knowledge-production process is a model of ontological reality and this includes scientific and non-scientific characterization of knowledge where mathematics and symbolic logic occupy different positions. The model of the ontological reality is the epistemic reality. In terms of scientific cognition, Max Planck illustrates the formation of model of primary reality in terms of world pictures. He states: *The scientific world picture or the so-called phenomenological world is not final and constant, but is in a process of constant change and improvement* [R14.87, p. 90]. The phenomenological world is what is referred to in this monograph as epistemic reality which is defined in the epistemological space. The epistemic reality is the work of information and decision-choice process through cognitive input-output processing machine where information is the input and knowledge is the output.

3.1 The Associations between the Primary and Derived Categories

In terms of associations between the primary category and the derived categories Max Planck states:

> *The first fact to claim our attention is that sensations, the sole and exclusive constituents of the world picture* [primary category] *have been driven appreciably into the background* [the derived categories]. *The dominant elements of this world picture* [derived category] *are not sensations but the objects which produce them* [R14.87, p. 94].

As it has been presented in these discussions to conceptualize the knowledge production through the knowledge square, we have the primary category defined in the *possibility space* through cognitive abstractions from the space of the *universal object set* (ontological elements) which is also called the *cognitive potential space*. It is a potential in the sense that it has a seed of knowledge that can geminate through work. The primary category is the input to our knowledge production system. Every aspect of the universal object set or the ontological

K.K. Dompere: Fuzziness and Found. of Exact and Inexact Sci., STUDFUZZ 290, pp. 35–51.
springerlink.com © Springer-Verlag Berlin Heidelberg 2013

space is composed of real and objective elements (composed of states, processes and things). These elements exist independently of human cognition. They satisfy the *identity principle* in the sense that they are what they are and exist in the potential space (ontological space) for knowing. The primary category whether empirically or axiomatically constructed contains the possible knowledge elements in the possibility space from which a scientific world picture (a set of derived categories) may be constructed on the basis of a societal culture of reasoning. These elements are obtained under defective information structure composed of vagueness, and limited information (from which propositions about knowledge are derived under linguistic vagueness and ambiguities in reasoning with inexact epistemic elements from the possibility set).

The elements in the universal object set in the potential space are taken to be exact in terms of what they are, and what characteristics define them. It is this mystery of what they are and what characteristics define their individual and collective existence, that epistemology is called upon to assist by first postulating the existence of a primary category. The primary category contains epistemic elements whose exactness and inexactness are decision-choice defined. Every aspect of knowledge is a simple cognitive picture. This picture is characterized by cognitive elements that are believed to exist by human decision-choice action on the basis of some justified conditions. It is then claimed to be certain, exact and knowledge possible in the possibility space by an individual or collective vote, at a time point, and stands for further epistemic works until it is revealed to be a phantom element by further collective decision-choice action; and even then, this decision-choice action is always temporary, awaiting for a new cognitive data that is supportive of conflict resolutions in the true-false duality and its logical continuum.

Every derived category that meets the true-false criterion of a decision-choice action, in the human cognition, is of a temporary nature. It sits on a dynamic vehicle of substitution-transformation process to assert the sequential motion among the derived categories where a categorial conversion of thought categories defines the evolution of stages of knowledge. This evolution is simply an enveloping of history of true-false decision-choice actions of human knowing. In this process on the path to knowledge, every derived category in the logical substitution-transformation chain is a parent of a derived category after it, as well as a child of a derive category before it by the method of constructionism. It is this chain of categorial conversions, with defined categorial moments, that provides an epistemic justification to the method of reductionism back to the primary category of knowing. The derived categories are equivalent to Max Planck's world pictures.

The dynamics of the categorial conversion between the primary and the secondary worlds are alternatively stated by Max Planck as:

> Every world picture is characterized by the real elements, of which it is composed. The real world of exact science, the scientific world picture, evolved from the real world of practical life. But even this world picture is not final, but changes all the time, step by step, with every advance of inquiry [R14.87, p. 96-97].

We must try to conceptualize the explications of Max Planck's world picture as seen in the initial and secondary and as it relates to the current statement of primary category and derived categories. The idea and position in these discussions are that there is nothing exact and certain about the derived categories except those that are decision-choice determined under irreducible core of epistemic uncertainty.

The world of exact science is a scientific world picture. It is formed from information from the epistemological space. It is, thus, a model of a representation that evolves from a primary world (category) while each previous world picture (derived category) molds the emerging picture, and the emerging scientific world picture presents a reflection of the future world pictures (subsequently derived categories). Any scientific world picture is exact to the extent to which the information input is exact and to the extent to which the logic of reasoning is also exact. Thus, cognition presents us with dynamics of logical substitution-transformation processes with the emergence of a new scientific world picture of epistemic reality and the epistemic disappearance of the previous scientific world picture of epistemic reality by displacement and substitution of parts or all. Each acceptance of a scientific world picture to be true, exact and certain is by decision-choice action in reconciling the conflicts in the true-false, exact-inexact and certain-uncertain dualities that operate under their respective internal dynamics in the logical continuum.

The logical continuum presents a thinking system where every truth contains falsehood as its support; every exactness contains inexact characteristics as its support and vice versa and every certainty contains characteristics of uncertainty as its support and vice versa, all of them operating as dualities which is equivalent to the *asantrofi-anoma* problem in the cost-benefit duality where every element is simultaneously good and evil [R2.9]. Thus, at each epoch of the knowledge construction process, we can specify the characteristics of the ruling scientific world picture of epistemic reality. The substitution-transformation process and the internal dynamics of true-false, certain-uncertain and inexact-exact dualities operating under conditions of methodological constructionism and reductionism and the principle of categorial conversion, have built-in corrective mechanisms in the knowledge-production process and the enterprise of science in a way that rejects a-priori exactness and certainty. By accepting a-priori exactness and certainty, one deprives us of the constructive-destructive creative dynamics of our knowledge production enterprise that allows affirmation and destruction of the current knowledge structure and the creation and substitutions of new additions.

The scientific world picture, composed of both primary category and derived categories, is not permanent and claims to its certainty and exactness can also not be permanent. The point here is that the claim of certainty and exactness of science is, perhaps, cognitive illusion where today's scientific truth may be tomorrow's scientific fiction both of which are products of our knowledge culture. This lack of permanence is stated by Max Planck as:

> *It must be noted that the continual displacement by no human whim or fad, but by an irresistible force----- Such a change becomes inevitable whenever scientific inquiry hits upon a new fact in nature for which the currently accepted world picture cannot account* [R14.87, p. 98].

The continual displacement of one world picture by another in search of an exact scientific knowledge may be viewed in terms of substitution-transformation processes among epistemic categories where there is always a category that is held as the primary one while others are held as derivatives thereof. The process requires updating and refinement of the elements in the primary category as new sense data appear on the epistemological space. It also requires a continual refinement of the thinking process as the existing derived categories are incapable of producing some knowledge accounts in terms of resolving some ignorance of interest. Each scientific world picture may be viewed as an epistemic category, and a movement from one picture to another is basically a categorial conversion by an epistemic substitution-transformation process that demands qualitative and quantitative laws of motion in propelling one scientific world picture to another.

3.2   The Evolution of Scientific World Picture from Inexactness to Exactness

The laws of motion that transport the scientific world pictures over the transient process are composite epistemic actions as established and collectively accepted by the knowledge search team. They are logical rules of categorial conversions that allow each scientific world picture to be examined, massaged and refined into other epistemic categories of scientific world pictures. These rules are cognitive creations to establish input-output epistemic technology that transforms information as input into knowledge as output. There is nothing absolutely exact or certain about these processes that help to define any knowledge sector as exact and certain. The epistemic elements of each world picture first reside in the possibility space as a possibility set then transformed into the probability space by a logical processing technology into a probability set for falsification, or verification, and then further transformed into an acceptance decision-choice action in the space of epistemic actual for corroboration with an element in the ontological actual. What can be claimed to be consistent and universal with the history of the enterprise of knowledge and in the space of scientific production is that of inexactness and uncertainty. Inexactness and uncertainty resulting from the defective information structure constitute the universal principle in all areas of knowledge production. Complete exactness and complete certainty are ultimate goals of knowledge production and the hard epistemic work is an asymptotic process to them. These ultimate states are such that we will never get there because the knowledge production is an infinite process just like the ontological elements are infinite whose ultimate knowledge is hidden from cognitive agents requiring continual search with epistemic tuning and retuning. It is here that the statement: *when one problem is solved, it gives rise to new problems* acquires meaning and epistemic beauty. Every solution creates a problem and every problem has a solution, all of which reside in duality. In this way, life is a never-ending process of problem-solving as cognitive enveloping that gives meaning to human existence in the space of inexactness. This is the basic central claim of principle of universality in the theory of the knowledge square as a meta-theory on the theory of knowledge. The solution path is the enveloping of the problem path and vice versa.

The inexactness and uncertainty composed of vagueness, ambiguities, instrumentation imprecision, and information limitation are such that our knowledge-production processes pass through fuzzy spaces with penumbral regions of decision-choice actions to select what is acceptable and what is not as a knowledge element or scientific truth. The epistemic substitution-transformation processes are basically fuzzy-tuning processes to reduce vagueness and increase degrees of exactness in concepts and reasoning within the fuzzy spaces for quality control under fuzzy-stochastic conditionality. Refinements in observations and measurements in terms of reducing the information limitations are, however, undertaken to reduce incompleteness in the sense data and increase the degrees of certainty as we move through the stochastic spaces. Explication is undertaken to reduce ambiguities in concepts, definitions and representations as we pass through the logical spaces and move through the penumbral regions of thought constructions. All these actions are elements of fuzzy-stochastic tuning of the epistemic process to reduce inexactness and uncertainties in the epistemological space. The implication here is that there is no exact epistemological space and hence there is no exact science that can be claimed under a-prior conditions.

All of these cognitive activities in the epistemological space are designed to create comparative relations between the elements in the epistemic reality and the elements in the ontological reality through the process of knowing by cognitive transformations. It is here that the logical strength of the theory of the knowledge square presents us with a universal principle that each knowledge search takes place in the epistemological space and passes through the connecting cords that link the four blocks on which the house of the epistemological space rests. The path of knowledge is thus an enveloping of epistemic transformations of world pictures of ontological elements. In these epistemic transformations, the characteristics of the previous world picture (or derived category) do not necessarily loose all their properties or disappear as knowledge elements. Some are retained, some fade into the background and some disappear into the potential. In fact, each category is defined by its own internal opposites in a duality whose inner conflicts provide the energy for the dynamics of categorial conversions among categories. Such knowledge transformations are gradual with information incompleteness, information vagueness, inexact representation and approximate reasoning methods within the continuum of dualities.

The movement from one epistemic category (world picture) to another may be understood from the analytical strength of the theory of the knowledge square. From the theory of the knowledge square two epistemological spaces are identified for the epistemic journey. There is the classical exact epistemological space characterized by information exactness with an exact input-output transformation technology and exact epistemic vehicle for the cognitive journey. Then, there is the fuzzy epistemological space characterized by defective information with an inexact input-output transformation technology and fuzzy epistemic vehicle for the cognitive journey. The claim against exact science is that all knowledge searches take place in the fuzzy epistemological space where information inexactness (vagueness, ambiguity and approximations) and information incompleteness are the rules in human thought, and accepted exactness and certainty are defined by decision-choice actions in the language of scientific work and subjective judgments. It is here that the possible worlds of logic and science meet the possibility theory from the fuzzy logic that takes us to the probable worlds and then to the space of epistemic actual. Over the fuzzy epistemological space, the cognitive journey allows subjectivities and ambiguities to be incorporated into human reasoning and decisions under various information constraints as we move through the penumbral regions in judgment of claims of degrees of exactness and certainty, which are human constructs under fuzzy and stochastic residuals that are expressed through the fuzzy and stochastic conditionalities. The concept of fuzzy-stochastic conditionality admits of vague and inexact probability in the knowledge systems. The relationships among methodological paths of the world scientific pictures, the primary and derived categories of knowing are shown in Figure 3.2.1. The projected relational path for methodological constructionism and reductionism in the knowledge production process is not knowledge-area dependent. Furthermore, the path is the same for the journey over either the classical exact epistemological or fuzzy epistemological spaces.

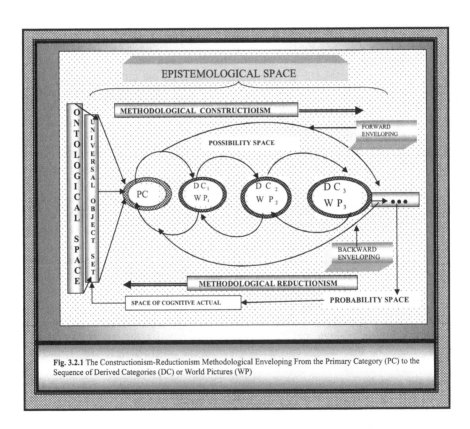

Fig. 3.2.1 The Constructionism-Reductionism Methodological Enveloping From the Primary Category (PC) to the Sequence of Derived Categories (DC) or World Pictures (WP)

In the process of cognitively categorial conversions, we project a transformational trajectory of non-linear continual improvements through adjustments, corrections of the epistemic world picture in terms of science and non-science by the principle of logical refinements through continual fuzzy tuning and stochastic tuning, where we move from a complete ignorance to a lesser and lesser ignorance such that the cognitive world picture becomes less muggy, less vague, less ambiguous, and less imprecise with increasing amplitude of epistemic exactness and accuracy, where the world scientific picture takes a definite form in cognition through the fuzzy and stochastic processes. The explanation of the process between poles under the inexact symbolism and fuzzy reasoning requires us to replace the classical *law of excluded middle* with the *fuzzy law of continuum* where every statement must meet the conditions of fuzzy restriction. The epistemic geometry of the path of the knowledge-construction process as a forward enveloping (constructionism) and as backward enveloping (reductionism) from the universal object set through the possibility space and probability space to the space of epistemic actual is presented in Figure 3.2.1.

Any scientist, even in the so-called exact science, will point out that the discovery of scientific truth goes through inexact process, passing through penumbral regions of decision-choice actions where there are incremental improvements of the epistemic clarity and certainty that bring hope and increasing cognitive light which will help to reduce the fuzzy and stochastic residuals as well as the areas of penumbral region of doubt and epistemic darkness. In other words, the whole of scientific research and the discovery of scientific truth is an inexact and uncertain process just as any other knowledge area. The duty of science is not to claim exactness and certainty of its area of knowledge search but to find exact and certainty equivalences with appropriate methods and logic, in order to report the discovery of the emerging scientific world picture or the derived category of epistemic reality. The claims of exactness and certainty of science are incompatible with the presence of risk, failure and disappointments which are the basic characteristics of vagueness, ambiguities, observation-measurement inefficiencies and limited information that produce a defective information structure which then constraints the human cognitive capacity in the practice of sciences. The knowledge production system, to which science is a part of, is a self-correcting one in an infinite domain for continual improvement in exactness and certainty. This self-correction character is simply due to the principle of defective information structure that characterizes the epistemological space for the knowledge search, as well as the capacity limitations of cognitive agents with regard to information processing and ambiguities in reasoning as cognitive agents travel over the fuzzy epistemological space.

The act of improving the cognitive world picture is a process through logical conversions of categories from the epistemic reality to ontological reality. The improved scientific world picture (cognitive derivative) that is gained and refined through experience and logical transformation is simply captured in the epistemic actual space or the phenomenological world of human experiences and cognition. The different cognitive world pictures bring to human understanding different possible worlds; all of them structured with modifications and perhaps where epistemic elements in the preceding scientific picture are replaced by collective decision-choice actions to assert the dominance of the new scientific world picture (or the new categorial derivative) in our knowledge production. Given the defective information structure and its representation, the degree of exactness and certainty of the scientific world picture will depend on the mode of reasoning and the type of epistemological space assumed.

The mode of reasoning defines the trajectory of the categorial moment in a given language, institutions and the culture of thought. A question, then, arises as to what logical paradigm will serve us better in the transitional dynamics among derived categories where the primary category is infested with the viruses of information incompleteness and vagueness, and when the actual human reasoning is colored with approximation of quality and quantity in order to resolve the problems of ambiguities, inexactness and uncertainty. Each scientific epoch presents us with a different but an improved scientific world picture or a derived category from the primary category or an epistemic model of the world through fuzzy and stochastic tunings since the knowledge system is self-correcting and self-integrating. Within each scientific world picture (or a derived category) there are active seeds waiting to geminate to give rise to other possible cognitive world pictures that may be possible by the type of the epistemic input-output transformation technology used. Within these world pictures simultaneously reside arsenals of destruction and tools of constructive refinements of the previously accepted scientific world picture (derived category) in continual conflicts of exactness and inexactness, and certainty and uncertainty in cognition as it relates to quality, quantity and time. In fact, it is these active seeds of destruction-construction process toward perfection, exactness and certainty that provide the rationalities for continual knowledge search without end. Any claim of a-prior exactness seeks to do away with the destruction-construction process of the knowledge enterprise and the accompanying technology of human creations.

In other words, there are qualitative and quantitative equations of motion for the integration of old scientific world order in pictures through backward logical motions and qualitative and quantitative equations of motion for the construction of new scientific world orders in pictures as replacement of the old. The former is related to the logical reductionism and the latter is connected to the logical constructionism. The two together generate logical duality that presents us with a continually unstable logical tension whose path defines a sequence of temporary epistemic equilibria for different areas of knowledge production and different scientific pictures. Each scientific world picture (or a derived category) represents a cognitive equilibrium of temporary nature. The path of the history of science, therefore, is an enveloping of *cognitive fuzzy equilibria* under a constant tension of internal dynamics of the knowledge equilibrium-disequilibrium duality in a continuum. Anything less than this, seems within the logical confines of fuzzy rationality as a distorted epistemic presentation of our knowledge production process, and of the process of scientific discovery, especially when the framework of the theory of the knowledge square is considered. The presence of inexactness due to fuzziness and the presence of uncertainty due to information limitation generate tensions among different scientific world pictures, and propel thoughts on the path of the knowledge-production process. The tension must be seen in terms of qualitative disposition as projected by fuzziness and quantitative disposition of information that is projected by incompleteness. The laws of motion among categories are established by paradigms of thought that create the conversional moments of forces of logical transformations.

The tension between the fading old scientific world order (the existing derived category) and the emerging new scientific world order (the new derived category) is the transformation-substitution power that defines the needed equations of motion for qualitative and quantitative processes of categorial conversions which in turn provide an inexhaustible source of an insatiable thirst for a new knowledge within the knowledge seekers on the terrain of thought. This terrain of thought, given the primary category as established by either empirical elements or axiomatic propositions, passes through the penumbral regions of cognitive activities. The penumbral regions defined by vagueness and ambiguities provide the inexhaustible sources of instability, incomplete thirst, and inexactness of information signals which constitute the sense data with never-ending approximate reasoning for their processing into new knowledge discoveries. All of which are locked in the foundations of non-mutual exclusivity where both qualitative and quantitative representations may be viewed in terms of linguistic numbers in continuum but not as the classical numbers in fundamental discreteness. It is useful, for mathematical and quantitative dispositions, to think that linguistic numbers constitute the primary category from which the category of symbolic numbers in their discrete forms became derived by decision-choice process to solve the problems of disagreements in linguistic counting. Alternatively, one must see the category of qualitative reasoning as the primary category from which the category of quantitative reasoning and analyses are logical derivatives. Similarly, the continuum (analog) is a primary category from which discrete (digital) is an epistemic derivative.

The elements in the universal object set exist only in interconnected mode where not only that the elements are interrelated, but the development of our sensory data is such that the elements can manifest one characteristic set in cognition through their information signals in certain respect, and not in the other respect over the epistemological space. The result is that one cannot decide definitely whether any attribute assigned is exact. This un-sureness becomes more complicated if the elements of scientific interests in the universal object set are qualitative processes or qualitative-quantitative processes where transitional dynamics may deny us of any certainty or exactness (example is social transformations) no matter how sharp are our laws of reasoning. Here, the classical laws of thought of exact reasoning can only help if we assume these elements of vagueness away from our representations and cognition and hold on to the classical extremes as reasonable approximations of the points in the continuum. These points of continuum are assigned discreteness and connected to each other by epistemic enveloping to obtain a cognitive model of ontological continuum.

3.3 Questionable Claims of Exactness and Certainty in Science

In all these discussions on the basis of the theory of the knowledge square, from the universal object set with human ignorance, through the possibility space with possibility sets, through the probability space with probability sets and to the space of the epistemic actual with knowledge, can we claim, under a complete cognitive comfort, the epistemic existence of an exact science with unquestionable rationality in the classical paradigm? The difficulties and complications that confront the seekers of exact science, in the knowledge production enterprise, are reflected on by Max Planck whose states:

> The fact that although we feel inevitably compelled to postulate the existence of real world in absolute sense, we can never fully comprehend its nature which, constitutes the irrational element which exact science can never shake off and the proud name 'exact science' must not be permitted to cause anybody to underestimate the significance of this element of irrationality [R`14.87, p. 106].

This element of failure to achieve absoluteness of existence of exact science must not be attributed to irrationality of behaviors of the elements in the universal object set in the ontological space. In fact, the behaviors of the elements in the universal object set are rational within the environment of their existence and substitution-transformation processes of nature and society. The elements and their behaviors meet the identity principles in that they are what they are and will be what they will be. The irrationality must be seen in terms of two factors that we have discussed as essential characters of the epistemological space. One factor is that the elements and categories in the epistemological space are characterized by defective information structure that robs the cognitive agents of exactness and certainty in the knowledge production process irrespective of whether one calls it science or inexact. The other factor is that the mode of exact representation and the exact classical laws of thought for the information processing are incompatible with the defective information structure of the epistemological space. In this respect, the classical paradigm, that provides the knowledge seeker with a mode of representations, rules of thought and corresponding mathematics in the epistemic process where exactness and certainty are the foundation of its knowledge construction and reduction processes, is irrationally incompatible with the processing of defective information structure. This irrationality is a logical one that is attached to exact symbolism and exact laws of thought which created the intense debate between the formalist and intuitionist schools of logic and mathematics. This irrationality vanishes when one changes the classical laws of thought as they have been discussed in [R2.9] [R14.97] [R14.100] [R14.6] [R14.14][R14.15]. The assumptions of the linguistic foundation of classical paradigm, the exact symbolic representations and the laws of thought are built on disturbing grounds that weaken the pillars of exact science and other areas of knowledge production that hold on to the strict classical paradigm for unshakable acceptance.

3.3.1 Exact Science, the Epistemological Space and Paradigms of Thought

The claim of existence of exact science and its development from the defective information structure in the epistemological space may be rescued if we can establish an inexact symbolic representation of the defective information structure and develop laws of thought to process it where exactness emerge as a thought process governed by decision-choice rationality of cognitive agents. The difficulties in sustaining claims of exact science in the knowledge-production process may be seen from the epistemological space. There are two epistemological spaces which are available in the knowledge-production process. The nature of the epistemological space claimed by the cognitive agents will affect the path of the knowledge search and the nature of exactness and certainty claimed for an area of the knowledge production. The epistemological spaces available to cognitive agents are inexact and exact epistemological spaces which set the direction for the development of the analytical symbolism, paradigms of thought and formal languages and claims of truth and knowledge. The relational structure of the exact epistemological space and the building blocks supporting the classical exact system of thought are presented in figure 3.3.1. In this classical system, the

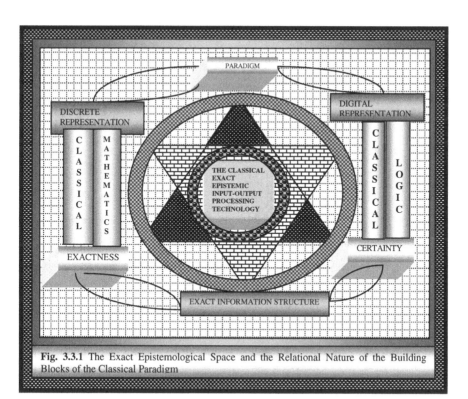

Fig. 3.3.1 The Exact Epistemological Space and the Relational Nature of the Building Blocks of the Classical Paradigm

defective information structure is due to quantitative incompleteness while it is stripped of conditions of vagueness, ambiguities and changing quality thus reducing subjectivity to irrelevant minimum to create conditions of exact topological spaces for the development of classical mathematics and logic. The possibility space is assumed to be exact that provides exact probability space for the refinement of the probabilistic epistemic elements. The exact probability space leads to the development of exact probability values with exact stochastic conditionality in all knowledge systems based on the classical paradigm. This includes exact science whose epistemic connectors are the principles of identity, opposite, dualism, excluded middle, leading to exact conclusions and certainty-values equivalences in logic and mathematics.

The relational structure of the inexact epistemological space and the building blocks supporting the fuzzy and intuitionist inexact system of thought are presented in figure 3.3.2. In this fuzzy system, the defective information structure is due to both quantitative incompleteness on one hand and conditions of vagueness, ambiguities and changing quality under the general concept of fuzziness on the other hand. The conditions of fuzziness introduce the importance of subjectivity into the knowledge-production process where decision-choice

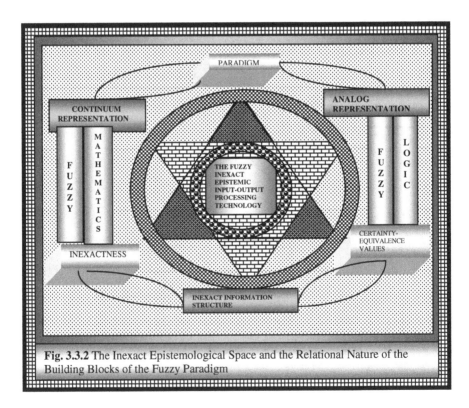

Fig. 3.3.2 The Inexact Epistemological Space and the Relational Nature of the Building Blocks of the Fuzzy Paradigm

actions are part of the knowledge production. The epistemic process is to create conditions of fuzzy topological spaces for the development of non-classical and fuzzy mathematics and logic. The possibility space is to be inexact which provides inexact probability space for the refinement of the possibilistic-probabilistic epistemic elements. The inexact probability space leads to the development of inexact probability values with fuzzy-stochastic conditionality in all knowledge systems based on the fuzzy paradigm. Thus, all areas of knowledge production including science proceed with epistemic connectors with the principles of identity, opposite, duality, continuum, leading to conclusions with exact-value and certainty-value equivalence.

The solution to the problems of representation of complete defective information structure and the development of appropriate epistemic operators for processing the needed information processing over the inexact epistemological space is the task of the fuzzy paradigm with its logic and mathematics which are in the speedy development path. It is this fuzzy solution path that will rescue exact from its critics. The exact classical epistemological space is put through fuzzification-process to reintroduce the conditions of defective information structure, and expand the epistemological space to include areas of penumbral regions of knowledge search. This allows the activation of the principles of duality and continuum to include the classical extremes. The new construct is then dealt with by a defuzzification-process to obtain fuzzy-stochastic conditionalities for conclusions and claims. We shall demonstrate the validity of the claim of the fuzzy paradigm as the path to rescue the exact science and its development from the defective information structure and to show the importance of the fuzzy paradigm in dealing with knowledge areas such as medical, social, biological and other sciences.

The lack of comparative appreciation of relational importance of inexact and exact sciences will be dismantled by the fuzzy paradigm. When exactness and inexactness are seen as linguistic variables and in degrees that reside in duality with an infinite continuum for the epistemological information about any ontological element, then the acceptance of exactness or exactitude as associated with any outcome of cognition or an epistemic process is human decision-choice based. Such results from the epistemic process are easily characterized as fuzzy rationality. The fuzzy rationality is decision-choice process by reconciling fuzzy set of exactness with its associated membership function and that of inexactness with the corresponding membership function by the method of fuzzy optimization. The result of fuzzy optimization leads to an optimal degree of exactness that yields an optimal value which constitutes the supporting acceptance principle of the level of exactness associated with the linguistic concept and cognition through a fuzzy decomposition decision-choice action. This fuzzy optimal value becomes the measure of *fuzzy conditionality* that must be interpreted in terms of the *fuzzy residual* in this monograph. Max Black calls it *the qualification of degree of consistency* [R19.4]. Günther Ludwig calls it, *the qualification of degree of uniformity* [R19.25]. The exactness and inexactness are partitioned and classified in terms of α-level sets under the fuzzy paradigm where every subject area whose exact membership value is greater than α^* is considered as exact with the corresponding degree of exactness qualified by a degree of inexactness that is tolerated. Similarly, every subject area of knowledge production whose exact membership value is less than, or equal to α^* is considered as inexact that is characterized with corresponding degree of inexactness and supported by a corresponding degree of exactness . In this respect, instead of the classical laws of thought and the classical rationality in the knowledge production, we have the fuzzy laws of thought and the fuzzy rationality in the knowledge production. The membership characteristic functions must be selected to fit the conditions of the principles of opposites, polarity and duality with fuzzy continuum where every true statement has a supporting false statement defined in characteristics recognition which may be seen in terms of cost-benefit configuration. Not all membership characteristic functions will meet the required conditions of duality and continuum. The membership characteristic functions must be continuous and inter-supportive of the negative and positive functional forms. The basic and conceptual structure will be discussed in Chapter 5 of this monograph.

At the level of a failure of absoluteness of exact science, Max Planck points out that:

> We sought a universal foundation on which to erect the edifice of exact science, a foundation of indisputable firmness and security – and we failed to find it...Our attempt was based on the idea of starting out on our scientific exploration from something irrevocably real, whereas we have now come to understand that such ultimate reality is of a metaphysical character and can never be completely known. This is the intrinsic reason which is doomed to failure every previous attempt to erect the edifice of exact science on a universal foundation valid a priori. [R14.87, p. 107].

Exact science, by its very definitional characteristics on the basis of individual experiment and data from experience, is subjective whose nature of exactness is decision-choice determined but not from any a-prior structure of human cognition. Exact and inexact sciences are defined and established in the epistemological space that is a creation of cognitive agents but not in the ontological space that is outside the control of cognitive agents. Its method of investigation can only be claimed to be exact by a cutoff decision-choice process that helps to place limitations and boundaries on the logical exactness and exactness of truth. What seems to be a *universal principle* that is being projected in this monograph for the erection of the edifice of exact science is that exactness is a derivative of inexactness by a process. Thus exact science is a derivative of inexact science where cognitive agents work with defective information structure to derive exact-value and certainty-value equivalences in the knowledge-production process through a decision-choice rationality. This universal principle applies to all knowledge sectors and will be related to quantity, quality and time that we have discussed earlier and will further strengthen the epistemic reflection. The whole method of exact science is a decision-choice process in search for the optimal path toward knowledge discoveries as we pass through the penumbral regions of epistemic challenges for clarity and certainty of what is claimed to be known about *what there is* or what is claimed to actualize *what ought to be*. The overcoming of these epistemic challenges requires the practice of methodological doubt but not methodological surety on the principle of universal foundation of certainty and exactness. The universal principle emerges from defective information structure of knowledge production where exact-value equivalence is derived from fuzziness of information and certainty-value equivalence is derived from information limitation. . [R11.22] [R14.100].

At the level of the classical thought, an attempt is made by Max Planck to rescue and rejuvenate the pillars of exact science from the reminiscence of the relevant foundations of classical paradigm. He states:

> *It is at this modest point (point of individual data of experience) that scientific research enters with its [exact methods] and it works its way step by step from the specific to the always more general. To this end, it must set and continually keep its sight on objective reality which it seeks, and in this sense exact science can never dispense with Reality in metaphysical sense of the term. But the real world of metaphysics is not the starting point, but the goal of all scientific endeavor, a beacon winking and showing the way from an inaccessibly remote distance* [R14.87, pp.107-108].

There are certain terms in the Max Planck's statement that requires some attention. His *objective reality* is the ontological reality while *the real world of metaphysics* is the exactness. All areas of knowledge production start with the perception of the elements in the ontological reality and exactness. The epistemic process in the epistemological space is to work from inexactness and create epistemic reality that is continually being refined judgment and decision-choice actions working with defective information structure.

At the level of cognition, the elements in the universal object set from the ontological space must be taken as absolute reality but unknown to humans. Exactness of human thought must not be viewed as the starting point of science or the knowledge-production process. Exactness of universal reality must be viewed as an ideal and a goal or a destination of all approaches of the knowledge production irrespective of the knowledge area of interest. Similarly, the exactness of methods and the logic of reason follow a destination path called the optimal path of knowledge production and should not be considered as a starting point of an a-prior nature. Human nature and its cognitive process toward perfection (exactness and certainty) from imperfections (inexactness and uncertainties) proceed in search of the perfect state as the available information space widens in intensity and in scope, and where an increasing learning takes place at each round of failure and each round of success. At the level of human cognition, exactness in science and hence all areas of the knowledge production, are driven by success-failure process with error-correction activities that tend to establish the path of history for the discovery of the optimality in logic and decision-choice rationality, as we work our way through epistemic complexities and uncertainties due to vagueness and limited information.

We may accept the notion that our individual experiences have been enhanced by complex physical instruments that have increased and continue to enhance the common agreement of our collective experiences as captured in the information structure, however, these physical instrumentations introduce different dimensions of inexactness. But this gain in instrumentation on the path of cognition is also the product of the success-failure cognitive process on the basis of inexact (approximate) reasoning and defective information structure as we take an epistemic journey through the penumbral regions of human cognition. This journey is a step-by-step or sometimes a movement on the ladder of scientific discovery and knowledge expansion. On the ladder of scientific progress and knowledge expansion, abstract concepts, logical samples and mathematics are used to help to reduce the fuzzy and stochastic residuals or the elements of inexactness in terms of *classical tuning* in an exact rationality. This process in classical paradigm tends to produce a thinking system at the top of the scientific ladder that becomes completely devoiced from the basic notion that the foundation of knowledge is a composite aggregate of individual and collective experiences that create defective information structure as an input into the epistemic process. The result is the creation of logical complexities that lose their bases and hide errors in reasoning and conclusions which, in the process, generate paradoxes and inexplicable contradictions. But the contradictions are parts of nature and form an essential characteristic of *what there is* and its identity. The instrumentation and measurements are tools for reconciling the conflicts in the quality-quantity duality.

This process of abstracting cognitive elevation under conditions of exactness from the primary category of reality through the sequence of derived categories at an increasing abstraction, where vagueness is abstracted out, further illustrates the inherent inexactness of all knowledge production that must pass through the penumbral regions of cognition. It is also here that the possibility space tends to emerge and also imagination arises in human judgment in moving from the universal object set (the cognitive potential space) to the probable, which will allow us to enter the space of epistemic actual which is simply a justified model of reality.

It may be emphasized that there are dangers to the efficient cognition and the knowledge discovery when an absolute claim is made to exactness of science as a special area of the knowledge production. A forceful plea to a general tolerance must be taken in unquestionable claims of truth as an essential element in moving our knowledge production slowly toward the best path of the knowledge acquisition irrespective of a particular area of knowledge. The optimal (most efficient) path of the knowledge production is the ultimate of which no one knows but assumes that this optimal path exists (see discussions in [R2.9]). Is the optimal path the same for all areas of knowledge production or is it different that varies over knowledge areas depending on the defined primary category of logical reality which is established by either empirical of axiomatic conditions? In other words, how does the optimal path depend on the logical starting point of knowledge production as it moves from the possibility space to the probability space and to the space of epistemic actual given the universal object set?

The efficient path of the knowledge production and discovery must be presupposed to exist for all areas of human logical endeavors and cognition. All knowledge areas must work in the epistemological space whose very nature is characterized by defective information structures that cannot be claimed to be a-priori exact. No area of the knowledge-production process can claim to have complete viewpoint of exactness and hence exact science. Perfect exactness associated with any area of human knowledge is an epistemic ultimate. It is a goal that logic propels seekers of knowledge toward. In this respect, it is useful to view exactness in degrees as defined by a fuzzy set and where complete inexactness and exactness are polar cases of living duality with a continuum under logical tensions that create conversion moments between them demanding decision-choice actions on the part of cognitive agents. These polar cases meet the Euler mini-max principle in universal substitution-transformation processes where nothing happens in the universe without defining minimum and maximum [R2.9]. The knowledge- production is the activities of both individual and the collective. It is also the work of information-decision-choice interactive process. These activities in the epistemological space require an organization to which we turn our attention.

Chapter 4
The Organization of Knowledge Construction and the Defense of Inexact Sciences

To understand the essential structure of the critique of the claim of exact science, which is being offered in this epistemic structure, it is useful to look at the organization of the knowledge-production process and the rationality of the knowledge acceptance in general. Knowledge construction is a problem-solving process that is executed through an organization, particularly when the problem solving involves two or more cognitive agents. The organization of the knowledge construction, in this respect, may be viewed as containing the organization of science whose focus is solving problems either through an empirical or axiomatic system of knowledge search and knowing. The task is to affirm or to change the set of existing conditions in support of what is claimed as knowledge or knowledge items. In other words, the organization of the knowledge construction is an integral part of the human problem-solving and decision-choice processes. The general human problem solving both at the level of knowing and otherwise, in turn, involves decision-choice processes where cognitive agents, with goals and objectives, are integral parts but not outside of the organization of knowing. The cognitive agents are the instruments of organization of knowing and the construction of the knowledge system. They are also the beneficiaries of the efficient functioning of the organization and the outputs of the knowledge system whether such outputs have negative or positive effect. Our contemporary scientific intellectual works and the knowledge production, on the aggregate and unlike previous periods, are extremely consciously collective activities organized to accomplish defined social goals and objectives. This does not exclude the unconscious part of the knowledge seeking process where positive accidents may work in favor of the knowledge seeking process. The knowledge-production process is a socio-economic activity of complex nature that works through the socio-economic boundaries of societies. These boundaries are established by the set of the institutions of economics, politics and law that shape the problem selection in terms of its relevance for the knowledge-production enterprise. The exactness of the knowledge production in general, and that of specific sectors of the knowledge enterprise will depend on the exactness of the creation and management of these institutions and the manner in which they are organized within the social setup.

K.K. Dompere: Fuzziness and Found. of Exact and Inexact Sci., STUDFUZZ 290, pp. 53–75.
springerlink.com © Springer-Verlag Berlin Heidelberg 2013

4.1   Social Institutions and the Organization of the Knowledge Production

Any social production, including knowledge, takes place within the confines of a society, where a society is viewed as an organization of individuals for multiple individual and social purposes. All these purposes relate to social production in the political, legal and economic structures whose behaviors determine the qualitative and quantitative dispositions of knowledge. The understanding of the quantity and quality of the knowledge accumulated in the knowledge house of the society and the additions that may be made to it, requires the understanding of the institutional processes through which researches for knowledge are made , claims to knowledge are presented and how the acceptances and rejections of knowledge items are subjected to decision-choice actions. The whole process of the knowledge production is a decision-choice process colored with total uncertainties composed of vagueness, ambiguities and limited information. The knowledge house of a society contains stock and net flow where such a net flow is an updating of the stock to expand the knowledge stock. The knowledge-production process, involving stocks and flows, in the decision-choice process moves in a direction that is governed by institutional rationality. The institutional rationality helps to define the environment of search, the selection of problems of interest, the methods of search, the channels of presenting findings, the criteria for knowledge claims, the methods of recording, the processes of justification and verification, and the methods of inclusion of items in the knowledge house. Alternatively stated, the knowledge process involves the knowledge-capital stock creation as a stock of intellectual capital and the knowledge net capital investment as an investment in the social intellect. The general structure of the knowledge production must be seen within the framework of an organization of society and its institutional framework for knowledge production in support of human existence and social advancement. It is, here, that the slogan the search for knowledge for knowledge sake loses it social appeal because knowledge production is not free. It demands the commitment of social resource in its production.

Our analytical focus, here, is on the role of social institutions for organizing decision-making for the knowledge production involving elements of nature and society. Additionally, the analytical focus is to examine and understand how the configuration of social institutions imposes on the knowledge-production process the quality of inexactness. Furthermore, how the quality of inexactness creates cognitive restrictions that can be fully understood with any reasoning that accounts for vagueness in linguistic systems, in inexactness of reasoning, ambiguities in conclusions and subjectivities in interpretations and knowledge acceptance. In this connection, it does not make much difference whether one is working in the category of explanatory science or the category of prescriptive science. The most important thing is the understanding of the organization of the knowledge production that allows the claims of exactness and the justification of such claims. It seems reasonable to expect that for an area of knowledge to claim

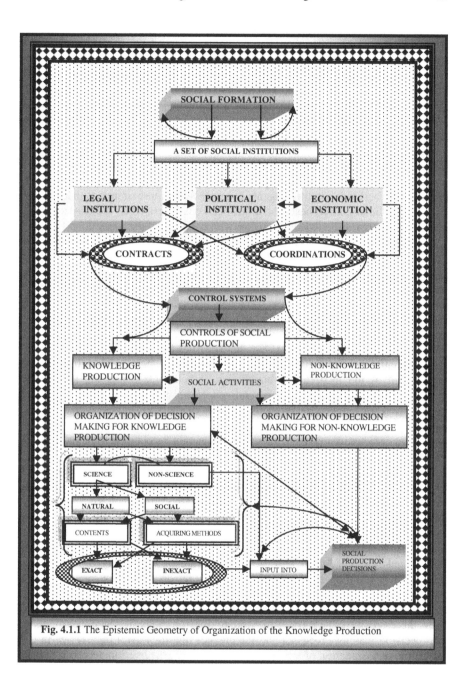

Fig. 4.1.1 The Epistemic Geometry of Organization of the Knowledge Production

exactness its organization of knowledge production must be exact. The exactness or inexactness of the organization of knowledge production resides in social science. One can have a claim to either explanatory or prescriptive science as being exact or inexact. The nature and characteristics of prescriptive and explanatory sciences have been offered in [R2.9] [R14.3O] and also [R16.14]. Whether the knowledge search is in the explanatory or prescriptive science, it follows the same path of epistemic geometry of knowing which is presented in Figure 4.1.1.

At the level of the nature of human action in the epistemological space, operating within the boundaries of cognitive limitations, expressed through a given language, defined within reasoning and decisions with defective information structure, the best that cognitive agents can expect are approximations that may be specified as the cognitive best which has been characterized as the classical sub-optimal rationality or bounded rationality with covering of the fuzzy conditionality in human general decision-choice actions when one subscribes to the concept of boundedness or levels of human aspirations in [R1.6] [R1.9] [R1.12] [R1.17] [R1.18] [R4.57] [R2.9]. All these cognitive limitations have been placed in the category of *fuzzy rationality* where human decision-choice actions are bounded by epistemic limitations imposed by vagueness of meaning, defective information structure and approximations in reasoning with language and symbolic representations that carry concepts, relations, thought and meanings. Decision-choice processes work with rationality that specifies the criteria of choice whether this rationality is implicitly or explicitly indicated. Such rationality is not independent of the institutional boundaries of held values and the knowledge seekers' preconceptions which also influence the problem selections and the methods of knowledge acquisition in all areas of the knowledge enterprise. The institutional boundaries of held values, culture of knowing and preconceptions are part of the defective knowledge structure and cannot be claimed to be exact as a-prior conditions.

Science is one of the areas of the categories of the knowledge systems. Exact science belongs to a category of knowledge systems. Scientist is a knowledge seeker in relation to sub-areas of interest and training. Such interests and training are institutionally imposed by the corresponding social value system and the resource constraints of general social production. All scientists, like other knowledge seekers, irrespective of the area of knowledge seeking, have values defined and imposed by the institutional structure of the society to which they belong. They would be devoiced of humanity if they do not have values within the institutional set up that specifies the allowable areas of search. They reason within a defined language that is used to represent information signals for processing. They, therefore, have the right of freedom to draw on the basic value judgments in their decision-making processes and the choices that they avail themselves in the search and determination of a knowledge discovery. This freedom to choose value judgments, from the institutional boundaries by the scientist, cannot be claimed to be neutral to science by ideologically obscuring them through abstract symbolic representations and claims of objectivity, exactness and certainty.

When the value-based propositions of scientific statements and culture of reasoning are made explicit, we quickly arrive at a conclusion that the claims of exactness of science and the objectivity in conclusions are conditional on the institutional value structure and the initial knowledge base from which preconceptions, vagueness, ambiguities, inexactness and ideologies are hidden. It is these hidden anomalies in the claimed exact sciences that define as well as encourage epistemic protective belts around scientific methods, justification and results of the scientific search. All the uses of abstract exact symbolic representations and the principle of exact reasoning cannot save us from these problems in science and the general knowledge construction on the basis of the classical paradigm with its logic and mathematics. For a symbolic representation without concept is knowledge illusory; concepts without linguistic representation are empty of knowledge.

The traditional system of scientific findings treats social framework, institutions and the enclosed values as given. In this respect, science becomes something abstract, objective and exact that projects restrictive views of scientific knowledge, scientific world picture and scientific discovery. By further imposing conditions of exactness, science becomes an entity that is not human operated and the scientist becomes emotionless like a mechanical machine that operates with logical rules of classical laws of thought in search of knowledge. The concept of exactness places science, particularly exact science, in a restricted epistemic domain where the findings of scientific knowledge become a system of rules devoid of institutional values and human judgment. In this way, abstract symbolic representations and logical operators drive the system of knowledge findings where there is no vagueness, and every word is claimed to be exact by definition, and where propositions are exact, and conclusions follow from objective rules without subjective judgment. In this way, the fuzzy residual and fuzzy risks that are due to vagueness in the information structure and ambiguities in thought in addition to subjectivity in the decision-choice actions are done away with. Additionally, the claim of certainty is to dispose of the risk of information incompleteness and sometimes discount the stochastic residual to be non-existent. This is not how our knowledge construction process seems to work.

At the level of cognitive being within the classical system, the scientist reduces himself or herself to the level of their instrument or to the level of a robot that operates with designed rules where such rules must be taken to be absolute. When a knowledge item goes counter to the designed rules it is ruled out as a contradiction and unacceptable or it is classified as a paradox in the system of thought. Of course this is not how scientists operate. They create and collect incomplete data, produce analyses, construct arguments, design justifications, make judgments, interpret results and communicate them. Any where on the path of the knowledge-production process that judgment is exercised, objectivity and exactness cannot be claimed unconditionally. The casualty in this *exactification* is the complexity over the epistemological space and where, triviality takes hold of the human mind, and where such triviality is justified under the principles of pure and exact sciences that confine imagination within the wall of *ridge rules* of thought which reduces cognition into simple epistemic algorithms such that we

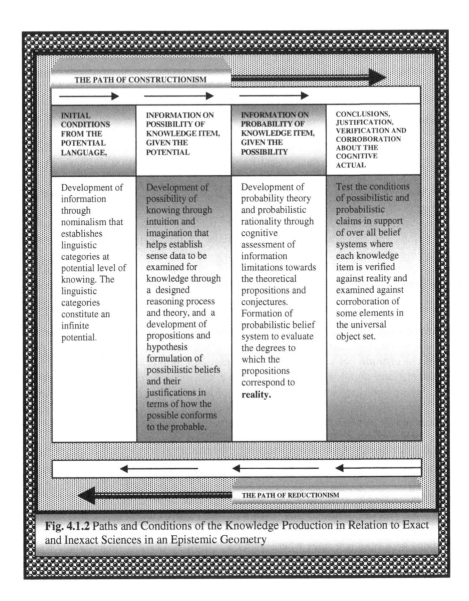

THE PATH OF CONSTRUCTIONISM			
INITIAL CONDITIONS FROM THE POTENTIAL LANGUAGE,	INFORMATION ON POSSIBILITY OF KNOWLEDGE ITEM, GIVEN THE POTENTIAL	INFORMATION ON PROBABILITY OF KNOWLEDGE ITEM, GIVEN THE POSSIBILITY	CONCLUSIONS, JUSTIFICATION, VERIFICATION AND CORROBORATION ABOUT THE COGNITIVE ACTUAL
Development of information through nominalism that establishes linguistic categories at potential level of knowing. The linguistic categories constitute an infinite potential.	Development of possibility of knowing through intuition and imagination that helps establish sense data to be examined for knowledge through a designed reasoning process and theory, and a development of propositions and hypothesis formulation of possibilistic beliefs and their justifications in terms of how the possible conforms to the probable.	Development of probability theory and probabilistic rationality through cognitive assessment of information limitations towards the theoretical propositions and conjectures. Formation of probabilistic belief system to evaluate the degrees to which the propositions correspond to **reality.**	Test the conditions of possibilistic and probabilistic claims in support of over all belief systems where each knowledge item is verified against reality and examined against corroboration of some elements in the universal object set.

THE PATH OF REDUCTIONISM

Fig. 4.1.2 Paths and Conditions of the Knowledge Production in Relation to Exact and Inexact Sciences in an Epistemic Geometry

forget that these simple epistemic algorithms are human created through decision-choice actions and subject to fuzzy and stochastic conditionalities as they relate to defective information structure and its processing.

It is useful to keep in mind that the decision-choice process encompasses all aspects of human endeavors that are enveloping of success-failure processes which generate epistemic risk in the knowledge-production system. It is this epistemic risk that is mapped onto decision-choice system as systemic risk of human action. All the knowledge-construction processes must account for risks of failure due to human factors and where vagueness and ambiguities define risk under assumption of exactness and even under information completeness. These human factors act through social institutions that constrain rationality in the ontological and epistemological decisions with the established boundaries of exact science and also inexact sciences. The institutional framework helps to define the conditions of *defective information structure* of fuzziness (composed of vagueness, ambiguities and subjectivity) and incompleteness of information. It has been pointed out that the former leads to fuzzy uncertainties that create possibilistic beliefs with corresponding fuzzy risks; while the latter leads to stochastic uncertainties that create probabilistic beliefs with corresponding stochastic risks. Given the institutional framework, we may view the path of the decision-choice process and defective information structure for knowledge production in steps as it shown in Figure 4.1.2.

4.2 The Choice of an Optimal Institutional Configuration

The best path of the knowledge-production process can be established by a collective choice in an institutional configuration from many possible paths that are constructible by human decision-choice actions. Some institutional configurations may be scientific-knowledge augmenting while some may be scientific-knowledge restricting. Such institutional configurations harbor collective beliefs and are not built by individuals. They are collective creations and evolve over time in a non-linear way. They help to define the quality of societies toward the knowledge production and technical capacity of the society towards further knowledge production. The choice of an institutional configuration establishes the framework of knowledge claims including exactness and inexactness in the scientific knowledge. Here, we observe that the knowledge stock performs a number of things. It is an input into further knowledge production as well as an input into institutional creation as a social capital. It is also an input into the decision-choice system and the management of the social system and its evolution. The institutions are the results of social formations whose behaviors are under the study by social sciences.

The knowledge output of the social sciences becomes an input into the knowledge production system. In this way, the exactness of a scientific knowledge is thus dependent on the degree to which the knowledge from the social sciences is exact. An increase in the exactness of knowledge in social sciences has important impact on societal ability to produce exact science which then affects the evolving quality of the social institutional configuration. Inexactness in the social sciences also transmits inexactness by extension to the other sciences and knowledge areas. The institutional configuration is complex. It is composed of entities, relations, interactions and mutual interdependence with conflicts which further aggravate the stochastic and fuzzy uncertainties in the epistemic space requiring a careful choice of the path for constructing the institutional configurations and their modifications.

Different institutional configurations of different societies have given rise to differences in knowledge accumulation and the speed of its updating. Differences in knowledge stock and the growth of knowledge may be explained by differences in institutional configurations and their contents of culture and value systems. The path of a choice of an institutional configuration including the modifications of existing one may be viewed as a dynamic social decision-choice problem and may be presented as in Figure 4.1.3 where $A_1, \cdots A_i, \cdots A_m, \cdots, i \in \aleph$,as an index set of socially constructible institutional configurations. An institutional configuration is a set of different institutions that may be classified under economic, political and legal structures. As viewed any institution organized to provide services will fall under economic production. In this way institutions of law and politics will play dual roles in the sense that they are institutions to regulate through the productions of services.

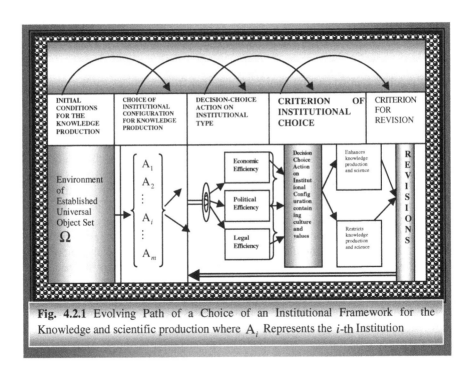

Fig. 4.2.1 Evolving Path of a Choice of an Institutional Framework for the Knowledge and scientific production where A_i Represents the i-th Institution

The choice of an institutional configuration that defines the framework of the knowledge-production process and other social activities is constantly evolving, establishing the dynamics of a science-non-science social intellectual game, in the belief justification and acceptance. The intellectual game is played in the integrated sectors of economics, politics and law. For example, we have science-religion game in determining socially relevant researchable problems, in evaluating conditions of acceptability of knowledge findings and the uses of such findings in assisting social management in general and further knowledge productions in particular. The evolving institutional configuration, when selected, establishes the dominant pattern of the cultural dynamics of thought in the knowledge enterprise. It further consolidates and protects the value structure of the boundaries of the knowledge production process through resource allocations on the basis of cost-benefit rationality whether the associated costs and benefits are measurable or not. The progress of the societal knowledge construction and the ability of the knowledge seeker to conquer new frontiers of knowing are ultimately governed by the quality of the institutional configuration that rules at any period of socioeconomic development. Both the institutional configuration and the knowledge-production system are interdependent and continually evolving providing inputs to each other and giving rise to a time-dependent institutional quality and cultural dynamics. The level of the quality of the institutional configuration is defined by the ruling culture, and the rate at which the level is changing is specified by the cultural dynamics of the social setup. From the view point of economic thinking, these are defined as cultural capital and cultural investment in the knowledge-production process. The exactness and certainty of any area of the knowledge production is a product of the interrelations of accumulated social infrastructure, cultural capital and knowledge capital. The progress of this exactness and certainty is at the mercy of the interrelations of infrastructure investment, cultural investment and knowledge investment.

4.3 Cultural Dynamics of Thought in the Knowledge Production and the Scientific Enterprise

The dynamics of thought relates to the evolving structure of the nature of thinking in natural and social sciences and in an interdependent mode. The dynamics of thought derives its force of qualitative and quantitative motion from the conflicts of what is claimed to be known and what is actually known in the practice of thought. In other words, it acts through a duality with an active continuum at any social time point. This evolving nature of thinking is institutionally defined, value-system dependent and shaped by the evolving and dominant culture. Alternatively viewed, there are creative interactions between the cultural capital and its dynamics on one hand and the knowledge capital and its dynamics on the other hand for any given social infrastructure. For example, the development of mathematics and the claim of exact science evolved from institutions of thinking in which the classical paradigm with its exact laws of thought was born. The principle of exact information structure and the laws of exactness of thought gave rise to the *laws of exact rigid determination* of true or false but not both that governed all areas of reasoning in the knowledge production such as the classical mathematics, classical mechanics, physics, symbolic logic and some areas of natural sciences. At the level of abstract thinking, the theories in these areas of the knowledge-production process about natural structures and behaviors acquired the character of *exact rigid determination* and where single-value solutions are attached to their relations such as equilibrium and stability solutions. These rigid laws of mutual determination came to characterize the knowledge acceptance by the ruling social institutions and culture of thinking for the knowledge production, especially in physical systems.

There are constant attempts to import them into other areas of knowledge production such as social, medical and biological sciences. These attempts have given rise to an increasing *mathematicalization* of different and complex areas of non-physical sciences and have produced some cross-fertilizations such as mathematical economics, bio-mathematics, mathematical social sciences, mathematical political science, mathematical psychology, computational science and many others which have culturally received greater social respect than their original subject areas without regard to irrelevance of some of the conclusions and the complexities as characteristics of the subject area of knowledge search where the laws of rigid determination from the classical paradigm are not applicable. The resulting process is such that an increasing emphasis is placed on mathematical and computational accuracy at the expense of conceptual accuracy with a complete neglect of the nature of the defective information structure for cognitive processing. This places the cognitive agents in the conflict zone of relevance and accuracy [R14.72] [R14.44][R14.75]. The laws of exact rigid determination are applicable to simple structures where qualitative complexities are absent or assumed to be irrelevant. The laws of exact rigid determination are influenced by Aristotelian laws of thought as characterized by the reasoning law of identity (where *what there is,* is *what there is,* but nothing else), the law of contradiction (nothing can be both *what there is* and *what there is not*) and finally the law of excluded middle (where everything must either be *what there is* or *what there is not*). These same laws of thought have come to influence the development of mathematical reasoning that allows the principle of exact-value determination to be applicable only to quantitative characteristics with neutrality of time between quantity and quality.

These laws of thought and the principle of exact-rigid determination have become engrained in some social thinking and the culture of the knowledge production as well as developed into an immutable ideology that is held on by scholars, knowledge seekers and scientists working in all the knowledge sectors whether their applications fit or not. The Aristotelian laws of thought deprive us of the existence of principles of duality and continuum where every element contains both of its opposites for identity. They further deprive us of the principle of conflict that brings about qualitative and quantitative motions and the changes from within the phenomenon as internal processes. The elements in the universal system are governed by internal processes where permanency is the characteristic of continual substitution-transformation processes that alter the relationships between quantities and qualities.

The Aristotelian laws of thought are consistent and applicable to working in the ontological space that is characterized by complete and exact information. Such non-defective information structure is not available to cognitive agents. If it were, there will be no need for knowledge search. Cognitive agents operate not in the ontological space of exactness of *what there is* but in the epistemological space of knowledge search and knowing. This space is characterized by a defective information structure that must be processed to know and explain the behavior of *what there is*. By applying the Aristotelian laws of thought to the knowledge search process in the epistemological space, one has assumed a one-to-one correspondence between the ontological elements to epistemological elements with exactness and certainty. In this way, the exact and complete information structure of the ontological space comes to replace the defective information structure of the epistemological space which then defeats the need for the knowledge construction. The only relevant assumption to connect the ontological space to the epistemological space is the assumption of the principle of identity where *what there is*, is *what there is* and the work of cognitive agents in the epistemological space is to find it.

The social and institutional acceptance of this method of exact single-value determination of quantitative characteristics is only useful when qualitative characteristics are assumed to be unchanging within a relevant period, such as physical particle or mechanical element or system in a defined time domain. And even here, we are confronted with the problem of vagueness that has been on continual discussion in physics and mathematics [R14.5] [R14.6] [R14.14] [R14.15] [R14.16] [R19.97] [R14.99] [15.23] [19.4] [R19.4] [R19.25] [R19.32] [R19.37] [R19.58]. In other words, the rigid exact-value determination demands extreme simplicity that allows a mechanistic view of the world to be developed with applications of classical logic and mathematics. When complexity increases with an increasing dimensionality, we are forced to deal with a defective information structure composed of vagueness, ambiguity and qualitative dispositions with the complexity. This defective information structure increases with increasing complexity to a point where the derived knowledge structure, on the basis of the classical paradigm, looses its meaning and application relevance to concepts. The rigid determination, in the exact science, is appealing because of the easiness of using higher and higher mathematical expressions to establish behavior where human value judgment, institutional constraints and subjective determination are suppressed in the knowledge production process. In this way, knowledge production acquires a mechanistic character that is removed from conditions of human action.

4.4 The Criteria for Partition of the Knowledge Space

The criterion for partition of the knowledge-construction enterprise into science and non-science and exact science and inexact science is developed on the basis of the mechanistic methods of knowledge acquisition. By accepting this paradigm of thought, we neglect the basic notion that knowledge production, however, is a human action and science cannot be exempted from this human action where such action, at all levels from the beginning to the end, is inseparably linked to human judgment that is guided by institutions and methods of social organization and management. The creation of these institutions and their social organizations are critically at the mercy of social science and the quality of its knowledge about the social organization and the relational and interactive processes of its contents. The development of exact science and the institutions for the production of exact science are at the mercy of the social science. In other words, the development of exact science and its methods of the knowledge production are social products and embedded in social science with all its limitations that are produced by ambiguities and vagueness leading to its characterization as inexact. The limitations of vagueness and ambiguities in human thought are essentially inseparable characteristics of all knowledge production systems and appear in varying degrees in different knowledge sectors depending on the complexities of the phenomenon in relation to *quality, quantity and time*. The vagueness and ambiguities increase when the system, like the social, biological and medical systems, must account for quality and continual quality changes that ensure their time-point identity of elements over the substitution-transformation trajectory.

When the complexity and dimensionality of a phenomenon increase which produce a further complication of increasing qualitative characteristics, the principles of exactness and rigid determination approaches to the knowledge construction are rendered inadequate or ineffectual. The increasing complexity and dimensionality create part-to-part interactions and interdependencies that provide the whole object with its quality and hence demand an account of a relationality of the parts and the quality of the internal organism that ensures its integrity. Here, as elsewhere, either the method of "inexactness" (imprecision) or approximate reasoning, that allows subjective relations to be studied as a problem of organization of the knowledge production, takes the central stage in such a way that an account is made of vagueness, inexactness, ambiguities and subjectivity in human thought and decision-choice action.

Here emerges the method of fuzzy paradigm to deal with the conditions of the defective information structure that characterizes the epistemological space and the activities of the knowledge search. It projects the search method of approximate reasoning or fuzzy logic that allows a conceptual system to be developed for dealing with the construction of an effective knowledge system to manage automatic-control and self-organizing systems such as society with constant institutional shifts that bring into being new qualitative dispositions and value changes. One of the greatest qualitative attributes of the knowledge-production process is that it is a self-correcting system with inputs of *defective information structures* on an infinite path. It may be kept in mind that exactness is not synonymous with certainty and truth. This method of approximate reasoning requires us to exit from the domain of logical dualism and abandon the classical law of *excluded middle* in reasoning that gives rise to *exact rigid determination* and then enter into the domain of duality, and replace the classical law of excluded middle with the fuzzy law of continuum in order to produce the conditions of *flexible determination* to replace the conditions of ridge determination where exactness and certainty are claimed as decision-choice action on the part of the cognitive agents.

The principles for organizing knowing that involves the phenomena of social, medical, biological and living systems with the use of methods of inexactness or approximate reasoning are intended to incorporate the *principle of exact rigid determination* as an extreme case in the *principle of inexact flexible determination.* Thus, we have available to us toolboxes of rigid and flexible computing technologies. Under these principles, the assumed information restrictions and the nature of information flows are essential in the knowledge constructions and the corresponding rationalities. The principle of exact rigid determination corresponds to exact information structure, classical paradigm with its laws of thought and classical decision-choice rationality. The principle of inexact flexible determination corresponds to defective information structure, fuzzy paradigm with its laws of thought and fuzzy decision-choice rationality where every area of the knowledge production is assumed to work from a defective information structure. It is here, and under the assumption of defective information structure as the input in the epistemological space and the use of vague symbolism that a methodological change from the principle of exact rigid determination becomes necessary. It is also here that a reexamination of accepted divisions of science into exact and inexact sciences calls into action a new rethinking of the knowledge production and its relationship to the acceptance of either the principle of exact rigid determination or the principle of inexact flexible determination with degrees of exactness that is decision-choice determined to produce a fuzzy conditionality.

By understanding the nature of the varying complexities of phenomena and the roles that are played by the possibilistic and probabilistic belief systems, with their corresponding measurements of fuzzy and stochastic uncertainties, an epistemic path may be opened to us to perfect the construction of theories and models of self-organizing, self-correcting and self-exciting systems where the exact rigid determination and uni-value representations create more difficulties in understanding relevance. This leads to a new approach for constructing scientific theories of exact and inexact sciences where probabilistic reasoning methods help us to deal with limited information in the defective information structure in scientific generalizations, and where possibilistic reasoning methods help us in dealing with vagueness, ambiguities and subjectivity in the defective information structure in scientific generalizations in the process of the knowledge construction.

The discussion, here, centers on the position that the general claim of exact science is artificial and can be humanized by relating it to decision-choice actions within the admissible confines of social and institutional set up. The exact rigid determination, as a methodological approach to the knowledge-production process, is limiting, especially when internal complexities, differentiations and dimensionalities in quality and quantity of the phenomenon of interest tend to increase. It fails to provide the user with a means to account for the two important deficiencies in cognition. These deficiencies are concepts of randomness and fuzziness in their scientific meanings. The randomness relates to information incompleteness (incomplete volume of quantitative information), the fuzziness relates to information deficiency due to quality of information in relation to vagueness in the linguistic representations, concept ambiguities, and subjective interpretations given the quantity of information. The information structure as an input into the cognitive processing for knowledge construction is such that fuzziness relates to incomplete *qualitative* information while randomness relates to incomplete *quantitative* information, thus, defective information structure is characterized by incomplete qualitative and quantitative information as an input into the thought process.

As it is presented, information that constitutes an input into the knowledge-construction process comes to us as duality of qualitative and quantitative information. The presence of qualitative disposition of information gives rise to exact-inexact duality while the presence of quantitative disposition of information gives rise to certainty-uncertainty duality. The fuzziness and randomness are essential attributes of every sector of the knowledge-production process where both possibilistic and probabilistic reasoning provide important cognitive flexibilities that are captured by the *fuzzy conditionality* and the *stochastic conditionality* to understand the internal structure of things and changes. For any aspect of science, therefore, to claim unconditional exactness, it must demonstrate the analytical methods by means of which fuzziness and randomness are qualitatively and quantitatively treated. In other words, it must show how quantity and quality of any information structure are treated in the epistemic process as well as to show how fuzzy and stochastic tunings are implemented as sub-processes over the epistemological space.

The potential space, with the universal object set, is an organism that is infinitely closed under the internal substitution-transformation processes where nothing is lost but continual quality and quantity conversions. The elements in the space are mutually and relationally held in flexible forms to allow categorial conversions that are consistent with the laws governing their internal structure for substitutability and change. The possibilistic and probabilistic methods of reasoning offer us a unified path of cognitive universality on the basis of human decision-choice action that allows subjectivity to enter into the internal structure of the knowledge-production process as human activity, and where quality and quantity of information characteristics are treated in a mutually organic process of knowing with neutrality of time.

The probabilistic reasoning, in compensating for incomplete information, increases the flexibility of thought and acceptance of conclusions on the basis of quantity of information available for the knowledge construction. We shall refer to this as *probabilistic flexibility* or *probabilistic inexactness* with *stochastic conditionality* in the epistemic process of human cognition. The possibilistic reasoning in compensating for information vagueness (approximate) increases the probabilistic flexibility of thought and acceptance of conclusions on the basis of the quality of information as determined by human action. We shall refer to this as *possibilistic flexibility* or *possibilitic inexactness* with *fuzzy conditionality* in the epistemic process of human cognition. The combined possibilistic and probabilistic flexibilities in cognition while reducing the importance of exactness, enlarges the range of cognitive accessibility of contemporary sciences and technology under critical investigation and analytical rigor through the use of the fuzzy paradigm with its mathematics and logic.

The analytical sum of possibilistic and probabilistic flexibilities in thought is the *epistemic flexibility* in the knowledge-production process. The sum of the fuzzy conditionality and the stochastic conditionality provides us with *epistemic conditionality* in thought. The advantage of this increased *epistemic flexibility* of cognition is the blurring of lines of demarcations between exact and inexact sciences as defined by the classical methods of knowledge inquiry. The importance of this is the extension of scientific and mathematical reasoning to areas of critical decisions in economic organization of enterprise of production, technological progress, medical services, behavioral sciences, physical experimental systems, knowledge of control systems, engineering systems, synegetics, informatics, energetics and complex social organizations. It is here, that the fuzzy paradigm with its logic and mathematics acquires methodological universality and rescues the inexact sciences from the attack of exact sciences where both quality and quantity of information are simultaneously treated.

The two reasoning methods with uncertainty for information processing under defective information structure have entered into the structure of scientific work and the enterprise of the knowledge production. The probabilistic reasoning, under the conditions of limited quantitative information, has gained a complete acceptance into the exact natural sciences as well as inexact areas, for example, of social sciences. In fact, this has become the standard way of analyzing and characterizing uncertainty and risk in decision-choice systems. In this way, the decision-choice actions with their success-failure processes are seen in terms of quantity but not quality of information. The possibilistic method of reasoning provides an approach of logic and mathematics from the fuzzy paradigm to enhance the probabilistic methods where acceptance of the degree of exactness and inexactness are human decision-choice determined. In this way, the reasoning within the fuzzy paradigm may come to enhance subjective probability measures and the likelihood process in the uncertain space. The importance of this, in the knowledge-production process, is that the fuzzy paradigm offers the cognitive agents a mathematical and computational approach to deal with the analysis of events in the fuzzy-stochastic spaces where inexact and vague or ambiguous probabilities confront thought [R15.8][R15.9] [R15.11] [R15.24]. Furthermore, the cognitive agents are provided with methods and techniques for providing vague symbolism and the laws of thought for processing the vague symbols.

4.5 The Possibility, Probability and Unified Epistemic Methods

The classical development of probabilistic system of thought, however, still retains the exact rigid determination which has become associated with the exact rigid probability determination under the classical laws of thought that rely on the classical frequency definitional approach to deal with quantity of information in its value determination. This approach of introducing epistemic flexibility in science by the method of non-fuzzy probabilistic reasoning deprives the knowledge seeker of subjective determination associated with ambiguities and inexactness of probability values that allow some flexibility to be introduced into the exact science where the conclusions are given their stochastic conditionality. The classical frequency principle is very limiting in constructing probabilistic reasoning when one considers the number of problems in the knowledge production for which frequency distribution is not available.

The approach through subjective probability or the Bayesian system also suffers from the use of the classical laws of exact rigid determination in the uncertainty space. The design of the multi-level probability measures (probability of probability) for the same event can not resolve the problem of quality of information such as vagueness and ambiguity. For example, problems of cybernetics, systemicity and general informatics in natural, medical and social sciences present fields of complex probabilistic systems that require determination of complex probability values that will account for vagueness and ambiguities in the characterization of probability values in all areas of sciences including the *exact sciences*. The manner in which the classical probability measure has been introduced into the scientific reasoning is mechanistic and seems to have nothing to do with human action and decision-choice determination in the information space. The theory of probability developed on the basis of classical laws of thought still follows the principle of exact-rigid determination without the account of qualitative characteristics of information. Alternatively viewed, the assumption of exact information structure is retained for the development of exact symbolism and the use of exact laws of thought under the conditions of incomplete information structure even though the attempt is to combine the cognitive agent and the decision agent into one agent, the human agent.

To introduce some degree of epistemic flexibility into the exact-rigid determination of classical probabilistic reasoning, the notion and the development of personal or subjective probability is brought into the laws of thought by the personal probability advocates. This is an acknowledgement of the fact that uncertainty due to insufficient information is a human limitation in the relevant space of the knowledge construction. Here, a different problem regarding subjective probability determination is encountered. The problem of the development of subjective probability is that the use of Aristotelian laws of thought does not help in escaping from the conditions of exact–rigid determination of the probability values. In both cases of the classical and personal probability determination, it becomes necessary to assume the existence of the possibility space from which a probability set is taken without which the axioms of probability will run into internal logical difficulties. In this process, the possibility set is taken to contain exact outcomes where vague outcomes are assumed completely away or taken to be non-existent. This amounts to assuming exact but incomplete information structure. The condition of exactness is then carried on to the probability space by a logical extension. In this way, we concentrate on exact probability space that allows the application of principles of exact-rigid determination and the classical laws of thought where the principle of excluded middle applies. Both the frequency and personal concepts of probability present us with some levels of interpretational difficulty. If A is a probability set then A is contained in P, the possibility space.

In order to develop the probability space, the possibility space must be assumed to exist. If the possibility space is not assumed but must be constructed, then the probability space acquires some extra complexity due to inexactness, vagueness and subjectivity. The method of abstract theoretical work on vague and inexact probabilities will benefit from the fuzzy logic and mathematics that allow the defective information structure containing vagueness and incomplete information to be simultaneously analyzed to produce a useful scientific thought that is driven by human decision-choice process which is subject to mistakes and faulty judgments and capacity limitations where corrections allow continual refinements and development of the scientific knowledge. The point that is being emphasized is that there is an important relationship between natural and social sciences and within the institutions that contain them. Defective information structure and both exactness and inexactness, just like certainty and uncertainty, are characteristics of all areas of the knowledge-production process, and our claim of one or the other to be exact is un-natural because the claim is human decision-choice determined under social and ideological conditions of thought.

The concepts of universal object set, possibility space, probability space and the space of cognitive actual introduced and discussed in [R2.9] [R2.10], as they relate to the natural and social sciences, are such that we are able to evolve techniques, methods and means to manage the society, nature and human thought. The interactions among society, nature and thought produce feedback and relational interactions in both the social and natural sciences where both areas of the sciences tend to affect their mutual progress. Thought draws its progress from society, society draws it progress from the practice of thought which is the parent of social and natural sciences and where the progress of exactness of natural science is dependent on the progress of exactness of social science in cognition. In this respect, we may ask a question as to the level of the degree of exactness and certainty contained in inexact science to provide us some level of epistemic comfort. Let us turn our attention to this question.

4.6 The Exactness and Certainty of Inexact Science

There have been a number of conceptual characterizations of inexact science. There are also a number of definitions at different levels of thought. The characterizations are structured on the relative basis of the defined boundaries of exact science which are then used to define the category of inexact sciences to include social sciences, behavioral sciences, medical sciences, biological sciences and sometimes applied physical sciences and engineering science. The defining characteristics depend on the criteria for distinguishing and defining the categories of sciences which we have already discussed. The elements of the required criteria for the demarcation of exact from inexact sciences are listed in Figure 4.6.1. The sets of criteria for separating exact from inexact sciences are the same sets that unite them into a unified process of the knowledge production. This unity finds expressions in the organization of the knowability and general knowledge construction about society and nature, in other words, *what there is* to be known, how is it known and many related things.

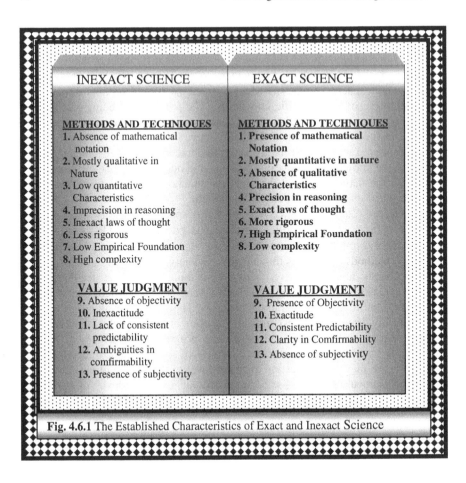

Fig. 4.6.1 The Established Characteristics of Exact and Inexact Science

4.7 The Role of Methology, Methods and Techniques in Distinguishing Categories of Sciences

The organization of knowledge production, irrespective of whether it is directed towards exact or inexact sciences, is done within a social set up and guided by decision-choice rationality in the reasoning process of cognitive agents in terms of allocations of social energies of resources and human-power. The allocations are done under social preferences over problems and questions where such preferences are shaped by culture of thought. Laws of thought are human creations and the uses of the laws of thought are human actions within the institutional and cultural confines of the social organism. Such organization of knowledge production is done through the existing and evolving social institutions whose behaviors are under the study by the social sciences. Now consider the universal object set, Ω and the organization of knowing. Every phenomenon, in the knowledge construction process, on the aggregate, has three dimensions of *quality*, *quantity* and *time* that define *quality-quantity-time space*. Elements in different categories are distinguished not by their quantitative dispositions but by their qualitative dispositions where changes in both quality and quantity have time dimension.   These three aggregate dimensions are characteristics of all natural and social phenomena. We may keep in mind that given any category, further classifications can be undertaken on the basis of quantitative disposition conditional on the qualitative disposition.

The quantity is easy to measure and *numericalized* with less disagreements when the units of measurement are agree upon, while the quality is not. This does not mean that measures cannot be found for qualities to create the needed conditions for *numericalization* as the science of measurement progresses, and the technology of measurement and instrumentation, all of which are human-action dependent, increases in complexity and accuracy. Both quantity and quality have subjective characteristics that must be accounted for in representations and in the knowledge-construction process. Both come to us through the senses which are the agencies of perceptions that empower the cognitive agents to construct information about quality, quantity and time. When time is taken to be constant through thought experiments, we then deal with epistemic static conditions of quality and quantity in the knowledge-construction process. This reduces the complexities of the knowledge-construction process and allows a critical examination of qualitative characteristics of the quantitative disposition of the phenomenon at a time point. On the other hand, when quality is held constant through thought experiments, this extremely reduces analytical complexities and the role that subjectivity plays in cognition. When quantity is held constant, the qualitative characteristics provide us with channels of formation of linguistic categories about the universal object set, and thus to distinguish one information item from another without which, the item of knowing is not verifiable or testable.

The relationship between qualitative and quantitative representations of a phenomenon is complex, so complex that care must be exercised in the knowledge production process. The relational complexity increases from physical phenomenon to biological phenomenon and to social phenomenon. It is this relational complexity that must be seen in discussions on exactness and inexactness of sciences where the existence of quality and quantity in duality helps to establish the identity of elements in the universal object set. It is also here that questions about exact and inexact symbolism become relevant as we relate them to the laws of thought in the knowledge-production process. Quality is a surrogate of quantitative disposition of any phenomenon whether social or natural or physical. It can be changed by the internal dynamics of the substitution-transformation process of the quantitative disposition. The converse holds in that quantity expresses the qualitative disposition of any phenomenon in the universal object set. It can also be altered by the internal dynamics of the substitution-transformation process of the qualitative disposition of the phenomenon in relation to time and the principle of categorial conversion.

4.7.1 The Concepts of Quality, Quantity and Time in the Knowledge Production

In the process of the knowledge production, the internal relational structure of quality and the quantity of a phenomenon in relation to time is vaguely reflected in thought and the nature of their representation is not only decision-choice determined but institutionally defined. The structure of simultaneous symbolic representations of quality and quantity in terms of languages and mathematics for reasoning will depend on the nature of their relational complexity in the phenomenon and the type of assumptions that is made to explicate this relationship. Depending on the phenomenon, one may choose to minimize the qualitative disposition or the quantitative disposition in the information representation for thought. The choice of representation of a phenomenon will determine whether we stick to the principle of *exact rigid determination* or that of *inexact flexible determination* while we keep in mind that exactness and inexactness constitute a duality as attributes of elements in the epistemological space. If one considers x as representing a phenomenon in the universal object set, then $x \in \Omega$ may generally be represented as:

$$x = f\left(\mathbb{X}, \mathbb{Q}, t\right)$$ (4.5.1.1)

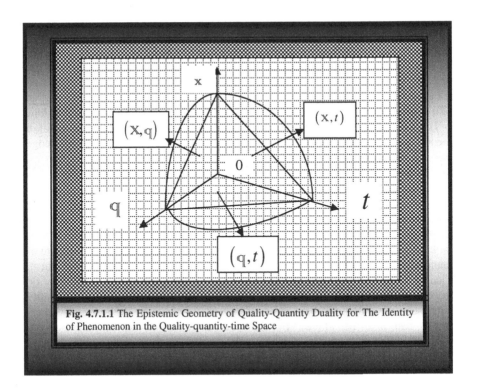

Fig. 4.7.1.1 The Epistemic Geometry of Quality-Quantity Duality for The Identity of Phenomenon in the Quality-quantity-time Space

Here, $\mathbb{X}$ is a set of quantitative characteristics, $\mathbb{Q}$ is a set of qualitative characteristics with aggregate value representation, $\mathtt{x}$ and $\mathtt{q}$ where t is the time element. The sets $\mathbb{X}$ and $\mathbb{Q}$ simultaneously establish the identity of x such that x is what it is at any time t, but its internal structure is under contradiction where such a contradiction may be assumed to be non-existing at a fixed time point as it is done in classical logical analytics and laws of thought. The nature of the epistemic geometry of quality-quantity duality and time is shown in Figure 4.5.1.1 in quality-quantity-time space.

The claim of exactness of a particular area of knowledge enterprise, such as science, is simply by assumption. In other words, exactness and inexactness are not ontological but are due to epistemic representations of cognitive reflections from the ontological space and the needed laws of thought which are human actions created within the framework of the institutional setup and the culture of thought. Inexactness in representation, from Figure 4.7.1 may be characterized by one or both of:

- a) Incompleteness of information in determining quantitative and qualitative nature of a phenomenon;
- b) Vagueness in the information of linguistic categories and symbolic representation of the qualitative and quantitative characteristics of the phenomenon;
- c) Indeterminate of the imprecision of time continuum.

Each phenomenon under epistemic scope cannot escape from these two elements of inexactness since they are generated by human action under human cognitive limitations in the epistemological space. The conditions a) and b) combine to define defective information structure that challenges representation and thought. In relation to the two conditions we may offer the following definitions for any given time set.

Definition 4.7.1.1

A symbolic variable in a representation of knowledge construction is defined to be mechanistic if it projects quantitative characteristic, neglects qualitative characteristic and is devoid of specific conceptual content.

Given the definition 4.7.1.1, we may write the variable representation in equation 4.7.1.1 as:

$$x = \begin{cases} f\left(\mathbb{X} \mid \mathtt{q}, t\right) = g\left(\mathbb{X} \mid t\right) \\ \text{and } \mathtt{q} = h\left(\mathbb{Q}, t\right) = \text{ constant, } \forall t \end{cases} \tag{4.7.1.2}$$

This is the approach adopted by Aristotle and the supporters of the classical laws of thought that allows quality-quantity contradiction to be disposed of as existing in categorial unity. In equation 4.7.2 quality is held constant and the quantity has no concept behind it. It is the same approach of representations that has come to characterize the classical mathematical systems and the criticisms by the intuitionist school of mathematics. The process is to obtain a representation of eqn. (4.7.1.2) where x is conceptually neutral to concepts and can be subjected to the classical laws of thought under the principle of non-admittance of contradiction in truth verification, and where the properties of the real number system hold under different operators. The path of isolating qualitative characteristics from the quantitative characteristics and to eliminate contradictions in any element in the universal object set is to cognitively and continuously dismantle the complex entity into micro-parts until the quality-quantity contradiction is eliminated in its duality such that x is x, and if

$$y = f\left(\mathbb{X}\mid_{q}, t\right) = g\left(\mathbb{X}\mid t\right)$$, then $y = x$ satisfying the law of identity. In this way,

complexity is removed to create simple forms where vagueness in thought, ambiguities in reason and subjectivity in interpretation are reduced to inessential minima that allow for only the analysis of the non-linguistic quantitative aspects without the part-to-part relationality of diverse forms. All linguistic numbers such as big, small, tall, fat, heavy, great, long, short and others that appear in human reasoning and communications are neglected. This approach to thinking is what is called *analytical thought* (*analytical logic*) that allows the classical laws of thought to be used with epistemic comfort because we only concentrate on micro-units without their organic ties to qualitative relation, synergetic nature of the phenomenon, by eliminating contradictions, excluding the logical middle and then focusing on the decision that is associated with the two extreme values of truth or falsity but not both.

The classical laws of thought, by arresting time and disposing of the essence of quality, have practically reduced the system of thinking to pure static states where, *what there is,* is *what there is* at all time points and hence inter-temporally universal. Even when the assumption on time is relaxed and allowed to flow, the conditions for the application of the laws of thought that are restricted to the extremes deprive thinking of its internal evolution due to the absence of contradictions and isolate the epistemic objects from the thinking individual who simply follows the classical laws of thought in manipulating abstract epistemic objects. This nature of the classical laws of thought makes it difficult to study substitution-transformation processes in self-exited, self-correcting and self-organizing systems, such as society, medical systems, environmental systems, the human body as a system and other areas of knowledge where qualitative characteristics are the essential elements of internal changes as time moves on and one category is transformed to another.

A problem arises when the object cannot be reduced to its component parts in terms of micro-units or when the analytical parts are to be synthesized into an organic whole where the different parts tend to produce interaction and some form of relationality to establish the *quality* that defines the identity and complexity of the whole. Here, contradiction cannot be eliminated and hence the classical laws of thought run into some logical problems in such a way that they cannot be preserved, especially the law of excluded middle. In this case, where complexity, dimensionality and relationality become multifaceted, producing vagueness, ambiguities, inexactness and subjective action, what then should guide thought or what form should the laws of thought take? Here, our emphasis is placed on the fuzzy paradigm.

The importance of the fuzzy paradigm is not to do away with quality and subjectivity in knowledge production, but how to incorporate them into the thinking process in order to derive useful and reasonable conclusions with the epistemic conditionality from defective information structures, where the epistemic conditionality indicates the risk of acceptance. The fuzzy logical component of the fuzzy paradigm helps to understand approximate reasoning, subjective judgment in inferential decisions, and relations of things to their defining qualities at the level of concrete existence to the abstract epistemic existence, and how rules of thought are subjectively defined by decision-choice actions under defective information structures. The fuzzy mathematical component of the fuzzy paradigm helps us to specify a computable epistemic system of representation of thought from defective information structure, on the basis of the fuzzy logic, through the reconciliation of contradictions in dualities, continuum and opposites. In this respect, the fuzzy logic and its mathematics are inseparably connected as well as connected to the continuum principle in dualities, polarities and opposites.

Chapter 5
The Exactness of Inexact Science and the Organization of the Knowledge Construction

The essential aspects of the organization of knowledge production as a human enterprise have been discussed in the previous chapter. In these discussions, the role of the quality of the societal knowledge on the social sciences is suggested to be an important constraint on the production of other sciences whether exact or otherwise. It is argued that there are epistemological differences, similarities and unity between exact and inexact sciences as categories of knowledge sectors.

5.1 A Reflection on Universal Principles of Ontology and Epistemology

The explanation of the differences is not attributed to what is generally held. The conceptual approach from the theory of the knowledge square, however, has a *universal principle of ontology* where, *what there is*, is *what there is* satisfying the identity condition at the level of existence. This universal ontology is subject area indifferent from epistemology and contains no inexactness. At the level of knowing, there is also the *universal principle of epistemology* where there is an abstract primary category and a series of derived categories whether the area of knowledge is classified as exact or inexact. As it has been presented, ontological objects are distinguished from epistemic objects. Both ontological and epistemological elements exist in categories of being and hence we can speak of ontological difference as well as epistemic difference. The epistemic difference is not necessarily the same as the ontological difference in that a horse and a goat may exhibit ontological difference in the sense of belonging to different ontological categories without exhibiting epistemic differences in a given language and thought system in the sense of belonging to the same epistemic category through knowing. It is here that epistemic errors are made of ontological elements to create a defective information structure where the epistemic knowledge deviates from the ontological knowledge about a phenomenon to give rise to a knowledge distance. The knowledge distance, therefore, is the difference between that which is claimed to be known and *what there is*. It is this distance that also gives rise to epistemic tests and the error-correction process to improve the epistemic knowledge in order to shorten the distance.

K.K. Dompere: Fuzziness and Found. of Exact and Inexact Sci., STUDFUZZ 290, pp. 77–101.
springerlink.com   © Springer-Verlag Berlin Heidelberg 2013

From the viewpoint of the knowledge-construction process and the transformations of epistemic categories from an epistemic primary to epistemic derivatives, the theory of the knowledge square considers categorial differences in epistemology as belonging to a grammar of thought that can and usually do produce cognitive differences and disagreements on the questions of what constitutes knowledge and whether such knowledge is exact or inexact because it is under the control of cognitive agents, who through decision-choice actions, must deal with as well as evaluate the qualitative characteristics of vagueness in language, ambiguities in thought, imprecision in measurements and approximations in reasoning. The ontological differences, as revealed in the epistemological space, belong to the grammar of existence that is not under the control of cognitive agents. At the level of epistemology, the methodological approach for abstracting knowledge items may vary over different knowledge sectors because of the contents and things that may be taken for granted by assumption and hence over the elements in the category of exact and inexact sciences. It has also been argued that all areas of knowledge sectors, no matter how they are formed, have overlapping boundaries and share in common the elements of information vagueness and limitation that generate inexactness of various kinds and degrees. All knowledge sector are united by a universal path to their production process but distinguished by the techniques and methods for processing the defective information structure whose constitution is knowledge-area specific. Let us reflect on the last statement by examining the relationships among language, vagueness and the knowledge system.

5.1.1 Language, Vagueness and the Knowledge System

The knowledge production, in general, cannot be isolated from a language and grammar of thought which by nature carry the characteristics of vagueness and ambiguities. The attribute of exactness and the degree of exactness that can be claimed, in this respect, are decision-choice determined and hence subjective in a significant way. In this respect, exact science, as claimed, has no undisputable foundation on the basis of which the exactness can be claimed with some full confidence. The characteristic set of exact science is imposed by simply getting rid of the quality attributes of *what there is* in the ontological space, as we move to establish an epistemological subspace that will be appropriate to the claims of the exact science within a given language in which we deal with analysis and reporting of analytical results in the knowledge subspace.

The knowledge-production system is carried by a language. The language is composed of *linguistic primitives* that are non-definables, *linguistic derivatives* that are definables and linguistic connectives that are transformation rules of thought. The relationships among the linguistic primitives, derivatives and exact science are under synthetic mercy of social sciences. Whether one justifies an empirical or axiomatic system or both as characterizing the primary category, the relationships are always meaningful within the language of their creation. It is within this language that quality, quantity and time become the epistemic focus of the knowledge construction and reduction. In this process, the meaning of any non-linguistic quantity is related to exact science without subjective interpretation, and when related to time, one deals with a process of quantitative changes with quality held as constant. This is basically a static system when one deals only in the quantity-time space. In the same light, the meaning of any linguistic term that is related to metaphysics is structured in terms of quality where subjective interpretation is required, and where the quantity may or may not be constant when time is referenced. In the quality-time space, one deals with the system's transformations and synergic processes. The situation is taken to define the characteristics of inexact science where qualitative dynamics and subjective interpretations enter into the space of knowing with the movement of the system, and where complexities are also the characteristics of the elements in the universal object set. The complexities are the results of the simultaneous interactions of quality, quantity and time, thus producing three-dimensional space for cognition.

In all sectors of the knowledge-production process, however, one cannot run away from the interactive relationships of the three basic characteristics of that which defines the elements in the universal object set. The three characteristics are quality, quantity and time as attributes in thought formation. In fact, quality as an attribute is an undisputable distinguishing characteristic that establishes identity, differences and categories of all ontological and epistemological objects. It is essentially, the qualitative attributes of elements, but not the quantitative attributes, that allow for the formation of linguistic, logical and mathematical categories that must be related to the development of knowledge. It is this attribute of quality that establishes the systems' synergetics and complexities. Water is different from milk on the basis of quality but not quantity or time. Hydrogen, oxygen and water are different through their qualitative characteristics even though both hydrogen and oxygen are in water. A unit volume of water is equal to a unit volume of milk in quantity but distinguished by quality which then affects their weights. The complexity of milk as established by its quality is not the same for water given the same unit of quantity and time. Water is different from fire because of quality but not quantity or time.

However, every quality that distinguishes one element from the other is basically a quantitative disposition where appropriate quantitative changes may lead to qualitative changes and hence changes in the essence of the element as it is being observed in time.

Every distinguishing quality of any element in the universal object set is the result of a critical organization of its internal characteristics to create a unique arrangement that presents a particular set of information signals to cognitive agents for its identity. Changing this arrangement changes its quality and complexity, and hence its identity, which then alters the application of the laws of thought and the conclusions that may emerge.

5.1.2 Vagueness, Knowledge Production and the Fuzzy Paradigm

The point that is being projected, here, is that the proponents of exact science have anchored their analytical power in quantitative accounts of the elements in the universal space. To dismantle the weakness of this grounding and its acceptance in the social setting, it is necessary to expose its foundation and its internal stresses that allow critical considerations of fuzzy paradigm and its laws of thought for alternative path of the knowledge-construction process. We are making an explicit epistemic point of entry into the epistemological space and methodological point of departure of the fuzzy paradigm into the knowledge construction process and the house of knowledge. Our objective is not on the developments of algorithms in the fuzzification and defuzzification of the classical system. The objective is to help establish the foundational need of the fuzzy paradigm in the knowledge construction in order to enhance the gains in the classical paradigm to include points of truth, false and contradiction in the classical excluded middle.

The important difference between the exact and inexact sciences is not established by their cognitive toolboxes or methodologies as viewed in the classical thought system but by the relative sets of their qualitative characteristics that are assumed to be contained in each element to assure its identity in the universal object set. In other words, exact and inexact sciences are partitioned into knowledge categories by the degrees of qualitative disposition that the informative object is assumed to contain for the epistemic process. It is the degree of quality in a phenomenon that produces varying internal complexities which in turn generates vagueness in representation, inaccurate measurements and observations, increasing difficulties in information recording and processing, approximations in reasoning, ambiguities in thought and the need for subjective judgment in various stages in the work of science.

At the level of knowing, the methodology as a philosophical category and its relevance as a global toolbox, for any given knowledge sector, is shaped by the degree of qualitative disposition allowed in the epistemic process, and this allowance is subjectively determined in terms of human decision-choice actions. Every phenomenon, under the process of thought, has irreducible quality which amplifies the irreducible vagueness and the irreducible fuzzy residual. The same irreducible quality contaminates the information sending-receiving process generating information limitations which magnify the irreducible stochastic uncertainty and the irreducible stochastic residual in the knowledge-production system. Here, methodology is conceived in a broad general framework that is different from the basic structure of epistemology given the ontology. The working definitional framework has been provided in Chapter one of this monograph. It is presented with the notion that the concept of epistemology is of less fundamental in the distinction between exact and inexact sciences, and among different knowledge sectors.

Given the ontological space, the epistemic process is initialized by the establishment of a primary category in the epistemological space where knowing produces cognitive movements from the possibility space to the probability space and then to the space of the actual. The cognitive inter-space movements demand different methodological vehicles between transient processes for different knowledge sectors due to differences in qualitative characteristics. Differences in the knowledge sectors arise from methods of investigation, methods of developing belief support, methods of reproducible justification, methods of reasoning, and the methods of knowledge acceptance as we move through the quality-quantity-time space.

As stated, the concept of methodology encompasses transient techniques, methods and specific reasoning in inter-categorial logical transformations with the constructs of justifications in both spaces of possibility and probability that allow the elements in the space of the cognitive actual to be established. The methodology, thus, defines the rationale on the basis of which the principle of acceptance of derived categories from the primary category constitutes rationality of knowledge elements. All knowledge areas are characterized by the same epistemic process of movements from the potential space that harbors the ontological elements, through the possibility space that allows the primary category to be constructed and the derived categories to be abstracted by an accepted mode of reasoning, into the probability space for acceptance test under probabilistic justification into the space of the cognitive reality. This is what the theory of the knowledge square presents as the process of content discovery where the methodological process provides us with the principles of content validation on the basis of socially acceptable decision-choice rationality in the knowledge production.

From the discussions in the previous chapters, the content discovery is plagued with uncertainties leading to inexactness due to a defective information structure, and this cannot be attributed to any specific area of the knowledge production. Since the presence of the defective knowledge structure is an attribute of all areas of the knowledge-production system, we do not have any undisputable assurance of explicating exact science except through human decision-choice action, and hence to establish the domain of inexact science by logical extension. Every area of the knowledge production must deal with the problem of inexactness in the primary category as well as with that of inexactness that characterizes *categorial conversions* processes over the defective information structure to obtain the derived categories or epistemic world pictures. It is this presence of the defective information structure as a constraint on perfection that leads Max Planck to the conclusion.

> *We sought a universal foundation on which to erect the edifice of exact science, a foundation of indisputable firmness and security – and we failed to find it...Our attempt was based on the idea of starting out on our scientific exploration from something irrevocably real, whereas we have now come to understand that such ultimate reality is of a metaphysical character and can never be completely known. This is the intrinsic reason which is doomed to failure every previous attempt to erect the edifice of exact science on a universal foundation valid a priori.* [R8.53, p. 107].

The term science, as a linguistic category is not a linguistic basic. It entails ambiguities in that it is not a member of linguistic primitive and hence has a definition or an explication. These problems of explication and definition are also true of exactness and inexactness in the epistemic discourse. The presence of defective information structure brings into the knowledge-construction process, fuzzy uncertainty, and fuzzy residual and fuzzy risk in the knowledge-acceptance process the understanding of which calls to action, the role of the fuzzy paradigm in all areas of knowledge production. The fuzzy paradigm defines a culture of knowledge production where contradictions are accepted and resolved in the knowledge production by decision-choice action. The culture of the knowledge production in this case defines a framework where non-exact symbolism is used and non-exact mathematical system is utilized to deal with quantity, quality and time to develop methodological *inexact flexible determination* with multiple values that flow from flexible computable system under fuzzy conditionality.

5.2   The Fuzzy Paradigm and the Explication of Exactness in Science

It has been argued from the structure of the knowledge square as presented in Chapter one of this monograph that the path of any knowledge production is characterized by possibilistic beliefs that relate to fuzzy uncertainties, and probabilistic beliefs that relate to stochastic uncertainties. The simultaneous presence of possibility and probability in the knowledge-production process presents us with the same symbolic, analytical, logical and mathematical difficulties that must be resolved in the epistemic process within any given formal language. The search for resolutions to these difficulties proceeds from the nature of the information sending-receiving mode, methods of information representation, reasoning, problem formulation and computing system, all of which constitute a paradigm of thought and knowledge production. Any selected paradigm must provide not only how to resolve these analytical difficulties, but must provide us with a decision-choice rationality for knowledge acceptance and rejection. We have discussed the epistemic problems of the classical paradigm, composed of its logic, mathematics and decision-choice rationality, in dealing with vague information and limited information that present us with total uncertainties in our knowledge structure. We shall now discuss the fuzzy paradigm, composed of its laws of thought, logic of reasoning, mathematics and fuzzy rationality, in order to examine the possibility of solutions to the epistemic problems in the classical paradigm at the level of information deficiency.

5.2.1   The Fuzzy Paradigm, Its Nature and Role over Inexact Epistemological Space

The essential characteristic of the classical paradigm that creates epistemic difficulties in dealing with conditions of fuzziness (vagueness, ambiguities and quality) in the knowledge-production system is its laws of thought with its excluded middle and non-acceptance of contradiction. This basic characteristic presents a logical dualism in epistemology. The classical paradigm, functioning with logical dualism and non-acceptance of contradictions, allows the exactness of science and classical mathematics to be justified where the principle of non-contradiction provides decision-choice rationality for true-false acceptance in the knowledge production process. But these conditions of the classical paradigm require us to do away with quality, subjectivity and the role of decision-choice agents as internal parts of the knowledge-production process. The knowledge seekers are thus externalized, with the effect that the knowledge-production system is stripped of the basic fact that cognitive agents are internally essential and integral parts of the knowledge production system. They are both producers and users of the knowledge.

The epistemic nature of the fuzzy paradigm and its development are a serious attempt to incorporate fuzziness, composed of linguistic quality, vagueness and ambiguities, whose understanding requires subjective decision-choice actions in the reasoning process to assess whether a contradiction is relevant or not in affecting the outcome. In this way, the individual subjective action is internalized into the knowledge-production enterprise as it should be. The cognitive agents are not simply machines that follow rules that are external to them, but they make the rules and execute judgments in an error-correcting process. This is in line with the position that the knowledge-production process requires value judgments that allow a traverse over the penumbral regions of decision-choice actions which involve inexact inductive and deductive thought processes in all areas of the knowledge-production process [R14.94]. In fact, the knowledge-production system is self-correcting and self-organizing. The fuzzy paradigm accepts the classical law of identity in the sense that ontological objects are what they are but rejects the classical position of dualism and the law of excluded middle. It accepts contradictions as part of human thought where such contradictions are resolved through reconciling the degrees of contradiction and non-contradiction in the logical unity through the decision-choice actions of cognitive agents under the principle of continuum.

To explain the epistemic process, through which contradictions are accepted in reasoning and resolved in thought, the classical *logical dualism* is replaced by *fuzzy logical duality*. This is an important fundamental change of the methodology of the knowledge-production process that will allow the inclusion of subjectivity, approximate reasoning and cognitive computing as internal processes of epistemology, where acknowledgement is made of limits of language in representation of information, derivation of thought, communication of ideas and interpretations of meanings. Even though the fuzzy paradigm is a revolution against the classical paradigm and also a contestant of the future paradigms, it actually enhances the subjective nature of thought in symbolic reasoning and widens the areas of mathematical and logical application in social, natural and other sciences. Our new scientific age, with information dominance, demands of us a new path of epistemic process and a new methodology of logical inquiry that will enhance our current methodological regimes in dealing with the understanding of both complexities and synergetics in phenomena. Here, it is interesting to consider new areas of informatics and energetics in our understanding of quantum phenomena at the level of nature and human behavior in relation to the mutual flows and conversions of information and energies into forces of change. In these discussions, we shall concentrate on the essential characteristics of the acceptance of duality and the rejection of dualism and their implied meanings in fuzzy paradigm.

5.3   Dualism, Duality and Unity in Cognition

A question arises as to why the classical dualism is being replaced with duality. Furthermore, is dualism not the same as duality, and if not, what are the distinguishing characteristics? In general, the concepts of duality and dualism have many different meanings involving morality, deity, ontology, mind-matter existence and many others where both of them represent a state of two opposite parts. In this discussion, however, we are interested in dualism and duality in paradigms of thought relative to epistemology and methodology of the process of knowing. Let us offer some working definitions by first considering the total characteristic set $\mathbb{C}$ on the basis of which the identity of an element is established in the universal object set, with two opposite sets of the negative characteristic set, $\mathbb{N}$ and the positive characteristic set $\mathbb{P}$, such that the relation $\mathbb{C} = \left(\mathbb{N} \cup \mathbb{P}\right)$ always holds.

Definition 5.3.1: Dualism

Dualism represents a conceptual state of mutually exclusive and collectively exhaustive two opposite characteristic sets that exist in the definition of the aggregate qualitative characteristic of a unit element without which the element has no defining identity in the universal object set. In this case $\mathbb{C} = \left(\mathbb{N} \cup \mathbb{P}\right)$ and $\left(\mathbb{N} \cap \mathbb{P}\right) = \varnothing$.

In terms of the Venn diagram this may be represented as in Figure 5.3.1.

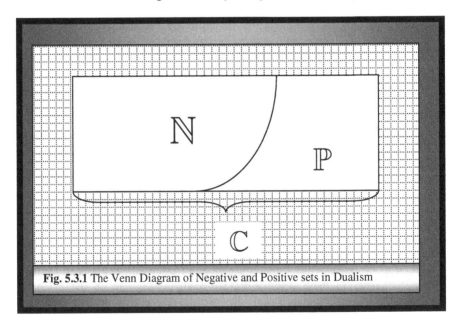

Fig. 5.3.1 The Venn Diagram of Negative and Positive sets in Dualism

Definition 5.3.2: Duality

Duality represents a conceptual state of two opposite non-mutually exclusive and collectively exhaustive characteristic sets that exist in the definition of the aggregate qualitative characteristic of a unit element without which the element has no defining identity in the universal object set. In this case $\mathbb{C} = (\mathbb{N} \cup \mathbb{P})$ and $(\mathbb{N} \cap \mathbb{P}) \neq \varnothing$.

In terms of the Venn diagram this is represented in Figure 5.3.2.

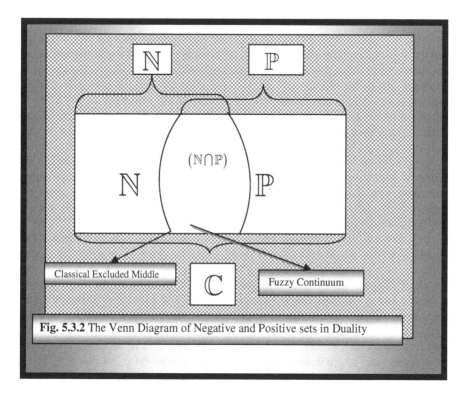

Fig. 5.3.2 The Venn Diagram of Negative and Positive sets in Duality

In the epistemic process and in examining the interactions of opposites we shall associate $\mathbb{N}$ with the set of false characteristics and $\mathbb{P}$ with the set of true characteristics. For the benefit of the discussions on exact and inexact sciences we could allow $\mathbb{P}$ to represent the characteristic set of exactness and $\mathbb{N}$ that of inexactness. In general, we can specify duality and dualism in set representations as in equations (5.3.1) and (5.3.2) respectively.

Duality implies that: $\mathbb{C} = \left\{ x \mid x \in \mathbb{N} \text{ or } x \in \mathbb{P} \text{ or } x \in \left(\mathbb{N} \cap \mathbb{P} \right) \neq \varnothing \right\}$ (5.3.1)

Dualism implies that: $\mathbb{C} = \left\{ x \mid x \in \mathbb{N} \text{ or } x \in \mathbb{P} \text{ and } x \notin \left(\mathbb{N} \cap \mathbb{P} \right) = \varnothing \right\}$ (5.3.2)

where x is a distinguishing characteristic relative to an element in the universal object set.

5.3.1   The Classical Paradigm and the Fuzzy Paradigm Compared

As defined, dualism and duality are different with a distinguishing characteristic that specifies a condition of mutual exclusivity such as false and true characteristic sets. In the knowledge-production process, dualism is associated with the classical paradigm and its laws of thought with *excluded middle* and the presence of non-contradiction, while duality is associated with the fuzzy paradigm and its laws of thought with *continuum* and acceptance of the presence of contradiction. In the classical paradigm, true and false are not allowed to simultaneously exist in the same proposition. In the fuzzy paradigm, true and false characteristics are allowed to be present in the same proposition where true or false is discounted by judgment with a fuzzy conditionality. Both paradigms have their corresponding mathematics and systems of logical symbolisms with different rules of mathematical and symbolic operators that may be used to generate and accept propositions and conclusions over the epistemological space.

The classical mathematical and symbolic operations function under the conditions of dualism and exact symbolism with exact laws of thought. The fuzzy mathematics and symbolic operations function under conditions of duality and non-exact (vague) symbolism with fuzzy laws of thought. For a comparative understanding, let us state side by side in Table 5.3.1, the two laws of thought that guide symbolic and mathematical operations. In the classical paradigm, words, propositions and conclusions are seen as points on the meaning and true-false lines where two opposing statements are seen in dualism and in separation with no connections, and where every statement is either true or false but not both. Here lies the principle of logical separation in the classical laws of thought on the basis of which propositions are accepted or rejected. In the fuzzy paradigm, words, propositions and conclusions are seen as sets on the meaning and true-false line where two opposing statements are seen in duality and in unity, each of which has sets of true and false characteristics that connect them in a continuum and in relative true-false proportions. Any statement is claimed to be either true or false by a decision-choice action with optimal fuzzy conditionality. Here, lies the logical unity of the fuzzy laws of thought in dealing with exactness and inexactness in the knowledge-production process on the basis of which propositions are accepted or rejected.

THE CLASSICAL PARADIGM	THE FUZZY PARADIGM
1. **The law of identity** (ontological condition) What there is, is, what there is.	1. **The law of identity** (ontological condition) What there is, is, what there is.
2. **The law of contradiction** (epistemological condition) Nothing can be and not be at the same time and space.	2. **The law of relative commonness:** or contradiction (epistemological condition) Everything is both what is and what is not in degrees.
3. **The law of excluded middle** (epistemological condition) Everything is either *there is* or *not there* is but not both.	3. **The law of continuum** (epistemological condition) Everything is in the process of being and not being. Alternatively, everything is both *what there is* and *what there is not* in degrees.
4. **The Principle of Separation** (epistemological condition) All opposites exist as separate entities.	4. **The Principle of unity** (epistemological condition) All elements exist in relational unity.

Table 5.3.1.1 The Essential Foundational Differences Between the Laws of Thoughts in the Classical and Fuzzy Paradigms.

The classical laws of thought, as applied to the acceptance of propositions and representations, satisfying the laws of identity, contradiction and excluded middle with dualism reduce to the statement –*all propositions are either true or false* [R14.100]. Comparatively, the fuzzy laws of thought reduce to the statement that *every proposition is a set of true-false characteristics in duality and continuum with varying relative proportional distribution and hence the acceptance of a proposition to be either true or false is obtained on the basis of subjective decision-choice action by cognitive agents in reconciling the conflicts in the true-false proportionality distribution in the continuum of true-false duality* [R2.9]. The reconciliation comes as a constrained cognitive optimization problem where the functional description over the true characteristic set is optimized with the constraint of the functional description over the false characteristic set. As applied to the classification of science, the exactness and inexactness exist in duality in continuum where the attainment of degree of exactness is constrained by the degree of inexactness.

The classical paradigm admits of absolutism in truth and falsity in the knowledge-production process. The fuzzy paradigm accepts the relativism of truth and falsity in the knowledge-production process. The epistemic importance and analytical interest of the two laws of thought are that the classical laws of thought taken as an aggregate statement cannot satisfy its own required condition of the statement: all propositions are either true or false but not both. This presents an epistemic dilemma in the classical system of thought. The fuzzy laws of thought can be subjected to its internal requirement in that its statement for guiding true-false acceptance is also defined with true-false duality in degrees and resolved through decision-choice actions by cognitive agents.

5.4 Duality, Continuum and Unity in the Fuzzy Laws of Thought

In the discussions on fuzzy laws of thought relative to the classical laws of thought as they may relate to the reasoning on exactness and inexactness in sciences and other areas of the knowledge-production process, we mentioned the concepts of duality, continuum and unity. It will be useful now to outline the essential characteristics of the concept of duality and how it presents a useful epistemic approach to the general knowledge-production system and a useful tool for reasoning in simple and complex systems under static and dynamic conditions. Complementing the concept of duality are the concepts of opposites, continuum and unity that help to explain and establish the nature of the fuzzy paradigm.

5.4.1 Duality, Continuum and Unity

In section 5.3, we examined the concepts of dualism and duality relative to logical unity and laws of thought. Their respective definitions were provided. We now want to examine in analytical details the reasoning foundations of the concept of duality in the fuzzy laws of thought. The concept of duality, like that of dualism, conceives of every element in the universal object set to exist in duality that holds two opposites which help to define its quality and identity. Every element in the ontological space is seen to be characterized by a set of characteristics in negative and positive standings, and where such characteristics are mapped onto the epistemological space for the knowledge-production process through cognitive operators that must be constructed. The ontological elements are related to the epistemological elements such as propositions, statements and conclusion and other relevant ones. For the identity of each element to be defined, the opposites must exist in unity in the epistemological space in an exact-inexact duality. The nature of the exact-inexact duality has been discussed in Chapter 4 of this monograph. The opposites are the extremes, defined by sets of the elemental characteristics that exist in a continuum without excluded middle in their residential duality, and where the points of indifference are cognitively computed by decision-choice actions under an accepted rationality and judgment.

To illustrate the point of discussion, let us take love-hate duality. Let us remember that in the fuzzy paradigm every element, word, concept, proposition and representations in their qualitative and quantitative dispositions are sets called fuzzy sets and the collection of identical fuzzy sets constitutes a category. Love and hate are fuzzy sets. For an individual, we define distributions of degrees of hate to which a feeling of hate is classified into the fuzzy set of hate, and degrees of love to which a feeling of love is classified into the fuzzy set of love in the same observational space in a spectrum. We must always keep track of the relationship between quality and quantity. Every quality can be expressed as a linguistic quantity whose comparative magnitudes are expressed in terms of linguistic numbers. A quality requires the specification of the characteristics that define the quality. For example, a set of characteristics of hate, love, beauty, good, tall, big and others are seen in terms of quality while their qualifying words provide the essence of comparison in linguistic quantity. For example, height may be specified as very short, medium short, average height, tall, medium tall, very tall, and many others. From the above, it is easy to understand the quality-quantity relationship in the logic of approximate reasoning. Every quality has quantitative disposition and every quantity has qualitative disposition where their meanings and understandings are expressed as sets but not points, and where every point of meaning has a fuzzy set covering. We speak of a number 4 and the degrees to which a set of numbers belongs to 4 such as close to 4 in a qualitative disposition. Another way of looking at this discussion is that linguistic numbers constitute the primary category from which the mathematical symbolic numbers emerged as derivatives. In fact, symbolic numbers may be viewed as explicated linguistic numbers.

For further clarification of the role of duality, continuum in the fuzzy laws of thought and how the fuzzy paradigm connects exact and inexact sciences in the epistemic process, let us examine Figure 5.4.1.1 that presents the relational structure of love-hate duality. The concept of love as a linguistic variable is represented as a fuzzy set, $\mathbb{L}$, with a generic element $\ell \in \mathbb{L}$, while hate, as a linguistic variable, is also represented as a fuzzy set, $\mathbb{H}$ with a generic element $h \in \mathbb{H}$, while the feelings of an individual may be considered as a set of characteristics, $\mathbb{F}$, with a generic element $f \in \mathbb{F}$, where $\mathbb{F} = (\mathbb{L} \cup \mathbb{H})$ and $(\mathbb{L} \cap \mathbb{H}) \neq \varnothing$. To specify the fuzzy set for any qualitative variable or linguistic variable that connects qualitative disposition to quantitative disposition, we define the membership characteristic function that specifies the degree to which an element belongs to the sets of both love and hate $\mu_{\mathbb{L}}(\bullet)$ and $\mu_{\mathbb{H}}(\bullet)$ respectively. The membership characteristic functions are more or less mappings from the qualitative input space to the quantitative output space while their inverses are mappings from the quantitative input space to the qualitative output. The geometric structures present the following fuzzy sets:

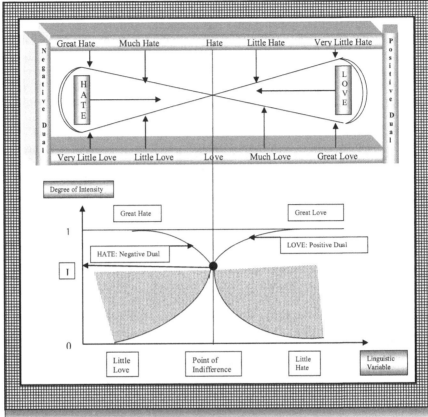

Fig. 5.4.1.1 An Epistemic geometry of Love-Hate Duality in Continuum and in Relation to the Fuzzy Laws of Thought and Approximate Reasoning

$$\tilde{\mathbb{H}} = \left\{ \left(h, \mu_{\mathbb{H}}(h) \right) \mid h \in \mathbb{H}, \mu_{\mathbb{H}}(h) \in [0,1] \text{ and } \frac{d\mu}{dh} \leq 0 \right\} \qquad (5.4.1.1)$$

$$\tilde{\mathbb{L}} = \left\{ \left(\ell, \mu_{\mathbb{L}}(h) \right) \mid \ell \in \mathbb{L}, \mu_{\mathbb{L}}(\ell) \in [0,1] \text{ and } \frac{d\mu}{d\ell} \geq 0 \right\} \qquad (5.4.1.2)$$

The two equations combine to form a fuzzy union $\tilde{\mathbb{F}}$ that defines a linguistic love-hate duality in the form:

$$\tilde{\mathbb{F}} = \left\{ \left(f, \mu_{\mathbb{F}}(f) \right) \mid f \in \mathbb{L}, \text{ or } f \in \mathbb{H} \text{ or } f \in \mathbb{L} \cap \mathbb{H}, \ni \mu_{\mathbb{F}}(f) = \left(\mu_{\mathbb{L}}(\ell) \vee \mu_{\mathbb{H}}(h) \right) \in [0,1] \right\} (5.4.1.3)$$

Instead of love and hate in Figure 5.4.1.1, one can substitute exactness and inexactness respectively and the thought process will still be maintained. The hate zone is where $\mu_{\mathbb{H}}(h) > \mu_{\mathbb{L}}(\ell)$ and the relative love zone is where $\mu_{\mathbb{L}}(\ell) > \mu_{\mathbb{H}}(h)$, and $\left(\mu_{\mathbb{L}}(f) - \mu_{\mathbb{H}}(h)\right)$ is the *fuzzy residual* in the love-hate space where the fuzzy residual is zero at the point of indifference with $\ell, h, f \in (\mathbb{L} \cup \mathbb{H})$ and $\ell, h, f \in (\mathbb{L} \cap \mathbb{H})$. We may also define the *fuzzy conditionality* around the point of indifference where $\left(\mu_{\mathbb{L}}(f) - \mu_{\mathbb{H}}(h)\right) = 0$ and $\left(\mu_{\mathbb{L}}\left(\ell^{*}\right) = \mu_{\mathbb{H}}\left(h^{*}\right)\right) = \mu_{\mathbb{H}}\left(f^{*}\right)$ in the form:

LOVE=$\left\{ l \mid \ell \in \mathbb{L} \cup \mathbb{H} \text{ and} \mu_{\mathbb{L}}(\ell) \in \left(\mu_{\mathbb{F}}(f^{*}), 1\right] \right\}$ Fuzzy conditionality for love   (5.4.1.4)

HATE=$\left\{ h \mid h \in \mathbb{L} \cup \mathbb{H} \text{ and} \mu_{\mathbb{H}}(\ell) \in \left[0, \mu_{\mathbb{F}}(f^{*})\right] \right\}$ Fuzzy conditionality for hate (5.4.1.5)

Examples of other dualities are good-bad duality, life-death duality hope-fear duality and many more that form part of the grammar of fuzzy reasoning. As it may now be clear, every antonym presents a duality in the language of its residence. With these examples of dualities as an illustrative background, let us return our attention to the main analytical concern regarding exact and inexact sciences.

5.4.2 The Characteristics of Duality in the Fuzzy Laws of Thought

Let us expand the analytical richness of the concept of duality and its distinction from dualism in relation to the fuzzy laws of thought. The essential defining characteristics of duality are:

1. Mutually interdependent opposites that are defined by characteristic sets which may be called positive and negative characteristic sets.
2. The opposites exist in a complex complementarity in that they mutually define the existence of each other in categories and categorial conversions without which each others existence is conceptually non-definable.
3. The mutual complementarity induces supplementarity, interdependence, reciprocity, in terms of give and take, and other categories defining qualitative and quantitative statics and dynamics of changes in relations such as independence, interdependence, identity and others.
4. Internal tension generated by the dynamics of internal conflicts between the sets of positive and negative characteristics in the process of accomplishing the basic relational characteristics specified in (1, 2 and 3) above.
5. Mutual negation that tends to change sameness into a difference and a difference into sameness, and the relative proportionality of the characteristic sets in a defined continuum of elements in terms of

categorial conversion in the dynamics of the substitution-transformation processes in thought. For further discussions see [R8.15].

These five statements are essential attributes in defining the concept and meaning of duality as distinct from dualism. The postulate of relational continuum may be linguistically related to quality and quantity of four points of qualitative and quantitative extremes. On the qualitative side, we have the extreme negative and the extreme positive. On the quantitative side of things, we have negative big and positive big. These extreme negative and positive qualitative characteristics and extreme negative and positive big define the identity of an element in unity under an epistemic tension and cognitive resolution. They are presented as an epistemic geometry in Figure 5.4.2.1 where there are movements from extreme positives to extreme negatives in the quality-quantity space. At the level of the primary category of the knowledge-construction process, the concepts of quality, quantity and time are epistemic fundamentals. These epistemic fundamentals are violated

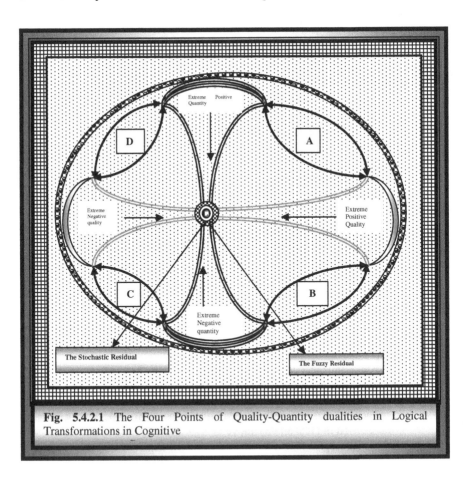

Fig. 5.4.2.1 The Four Points of Quality-Quantity dualities in Logical Transformations in Cognitive

by the acceptance of linguistically quantitative purity with exact measurements and mathematical representation that fails to account for the role of quality and subjective judgment in the quantitative positions. The qualitative characteristics in human experiential manifolds must guide us in the development of mathematical and logical reflections in representation of reasoning and geometry of thinking in reducing complexities into simpler forms towards irreducibility.

The extreme positive and negative qualities may be associated with complete exactness and complete inexactness respectively as the poles, and with the logic of excluded middle they are transformed into dualism where they do not share elements of their transformations. The exactness and inexactness may be applied to quantitative characteristics as what we know to be true. This position is not consistent with actual world of operations where the concept of exactness can only exist and be understood with the existence of the concept of inexactness and vice versa. The complete inexactness and complete exactness are conceptual poles of quality and quantity. The acceptance of exactness and inexactness comes with supporting subjective belief values which the fuzzy laws of thought advocate. In the classical laws of thought, opposites are recognized, but they are viewed as mutually exclusive and collectively exhaustive with nothing in common, where the membership characteristic function assumes the value of either 0 or 1 with nothing in between.

The connection of positive to negative through a give-and-take process and mutual negation as expressions of the path linking the complete exactness and complete inexactness must be done in such a way that we abandon the law of excluded middle and the non-acceptance of contradiction that form the conditions of the true-false acceptance in the classical paradigm. The classical laws of thought are also incompatible with substitution-transformation processes of mutual negations that allow the study of dynamics of quality-quantity changes in a specified time domain. It is here that the fuzzy laws of thought that may be derived from the conditions of fuzzy paradigm, as presented in Table 5.3.1, enter.

First, we observe that the law of identity establishes the ontological elements in the universal object set where the elements are free from the conceptual notions of exactness and inexactness because they are what they are and would be what they would be. Given the ontological elements, *what there is*, is brought to the epistemological sphere for examination, study and understanding. The laws of duality of commonness, unity and continuum (non-excluded middle) advance rules of reasoning and criteria of acceptance of contradictions where every element is and is not, and every element is always on the trajectory of being or becoming and not being or not becoming. The fuzzy laws of thought establish epistemic conditions for the acceptance and non-acceptance of the knowledge propositions, and where each acceptance has a fuzzy conditionality. Thus, it helps to establish approximate-reasoning tools to navigate through the penumbral regions of decision-choice actions between the positive and the negative extremes such as complete exactness and complete inexactness or absolute truth and absolute falsity in thinking and in the knowledge production process, as well as in any epistemic duality with defective information structure.

The conditions of the epistemic duality with the law of continuum in reasoning suggest that there is a process of interaction between sets of true and false characteristics that produces tensions which is resolved by defined decision-choice rationality as a utility tool for the cognitive agents. Since tension involves two tendencies of falsity and truth, it also involves two epistemic directions where acceptance of truth of proposition suggests that the truth is accepted to contain more truth characteristics than false characteristics. Another way of saying this is that the membership characteristic function of the truth characteristic line lies above that of false characteristic line as it is exemplified in Figure 5.4.1.1.

The fuzzy laws of thought, given the fuzzy paradigm that encompasses ontological and epistemic conditions, may simply be stated again as:

> *Every proposition contains true and false characteristics in varying proportions in such a way that true and false statements are determined by decision-choice action in resolving the conflict between truth and falsity, where the truth characteristic set is optimized subject to the false characteristic set, to provide a measure of fuzzy conditionality that is used to decompose the set into true and false statements for choice.*

This is an acknowledgment of the role that judgment plays as an important tool that is used by knowledge agents in the knowledge-production process and that inductive and deductive reasoning are constrained by defective information and cognitive limitations. In the fuzzy system of the theory of knowledge, every claimed knowledge item, therefore, cannot be definitive. Any claim of knowledge can be accorded varying degrees of validity for specified applicable domain with both stochastic and fuzzy risks defining the relevant domain. The stochastic risk in the knowledge production and human action is generated by the stochastic residual whose measure provides the stochastic conditionality while the fuzzy risk is generated by the fuzzy residual whose measure provides the fuzzy conditionality for every claimed knowledge item. The sum of fuzzy and stochastic risks presents us with a total risk of inexactness contained in the region of acceptance of truth or falsity and in the space of human action. All these operate in the cost-benefit space.

5.4.3   Duality, Continuum, Cost-Benefit Rationality and the Fuzzy Paradigm

Duality is a fundamental concept in all social and natural production processes. It manifests itself in both the ontological space and the epistemic space. Another interesting way of viewing the fuzzy laws of thought is through the analytic foundations of cost-benefit rationality as a general guideline for decision-choice action [R16.8], [R16.9]. Every decision is composed of a set of decision-choice characteristics, $\mathbb{D}$, with a generic element $\mathbf{d} \in \mathbb{D}$. The decision-choice set of characteristics exists in a duality of a set of cost characteristics $\mathbb{C}$ with a generic element $\mathbf{C} \in \mathbb{C}$ and a set of benefit characteristics, $\mathbb{B}$, with a generic element

$\mathbf{b} \in \mathbb{B}$ such that $\mathbb{D} = \mathbb{B} \cup \mathbb{C}$. The concept of continuum is brought into the cost-benefit rationality in that the benefits of every decision have their cost supports and the costs of every decision have their benefit supports where every cost characteristic may reverse its role to be a benefit, and every benefit characteristic may reverse its role to be a cost in a rational foundation of thought and decision-choice action.

The sets of the cost and benefit characteristics exist as inter-supportive and non-mutually exclusive opposites in any defined action of cognitive agents. They exist in an inseparable unity of a continuum in the cost-benefit duality with give-and-take relations but not in a dualism with mutually exclusive opposites without a relational sharing. In this way, $\mathbb{C} \cap \mathbb{B} \neq \varnothing$, and the decision-choice action $\mathbf{d} \in (\mathbb{B} \cap \mathbb{C})$ defines a constrained decision-choice action. The decision-choice actions are determined in the system of reasoning where the choice of the benefit characteristic set, associated with any decision-choice element, is constrained by the cost characteristic set contained in the same element in the cognitive calculus. The nature of the cost-benefit rationality and the general decision-choice actions under fuzzy laws of thought may be viewed in relation to general decision-choice rationality. The applications of the cost-benefit rationality under conditions of fuzziness in the theory of knowledge-production process are also discussed in [R2.9] [R11.21].

In this explanatory process of exactness and inexactness, one may ask a question: why are costs and benefits being introduced into the discussion of the fuzzy laws of thought? The answer to this question relates to a rationale which is direct but complex. First, there is a primitive proposition that every human activity or action is decision-choice determined under defective information constraint. Second, every decision-choice action or activity is determined by cost-benefit balances that may be abstracted from the activity. The knowledge production is a human activity. The fuzzy laws of thought provide a path of a decision-choice action on true-false acceptance in the space of the knowledge production. The true-false duality may be seen also as benefit-cost duality. Viewed in terms of the benefit-cost relation, the truth characteristic set may be associated with the benefit characteristic set and the false characteristic set may be associated with the cost characteristic set. The decision-choice actions in the acceptance of true or false propositions or knowledge items in the knowledge-production process are determined in a system of reasoning where the acceptance of a true proposition finds its justification under the condition where the truth characteristic set, associated with any decision-choice element, is constrained by the false characteristic set contained in the knowledge element. Thus, the true-false balances reduce to benefit-cost balances in the fuzzy laws of thought where every true proposition has its false support and every false proposition has its truth support. To understand the working mechanism of the fuzzy laws of thought one may consult any work on fuzzy symbolic logic [R3] [R3.2] [R3.26] [R3.27] [R3.28] [R3.45] [R3.52] [R3.57]. Again, one must see a reduction in the epistemic risk as a benefit while an increase in epistemic risk as a cost, all of which must be related to the resource allocation in the social knowledge-production system.

Every decision-choice alternative, in the human decision-choice space, has either an irreducible cost for an accepted decision or an irreducible benefit for a rejected decision.   As discussed, every duality, such as inexactness-exactness duality, is reducible to a cost-benefit duality that provides a rational foundation of decision-choice actions by the assessments of the negative-positive balances where the positive set is constrained by the negative set and vice versa in the decision-choice space. The decision-choice space in these discussions is conceived in the most general space of thought about human actions in every respect, including the knowledge-production activities. The cost and benefit characteristic sets are likewise conceived where they may or may not be measurable. The conditions of non-measurability are mostly related to qualitative characteristics of the relevant variables while the conditions of measurability are mostly related to quantitative characteristics of relevant variables. The information on the non-measurability in the calculus of decision-choice actions are represented by linguistic variables such as the ones in Figure 5.4.1.1. The approximations of the conditions of measurability in the calculus of decision-choice action are also represented as linguistic variables and linguistic numbers.

5.5   Algebra of Fuzzy Reasoning in the Inexact Science

Let us revisit Figure 5.4.1.1 where $\mathbb{F}$ is the characteristic set of feelings with a fuzzy set $\tilde{\mathbb{F}}$, the characteristic set of love $\mathbb{L}$, with a corresponding fuzzy set $\tilde{\mathbb{L}}$ and the characteristic set of hate is $\mathbb{H}$ with a corresponding fuzzy set $\tilde{\mathbb{H}}$ as specified in eqns.(5.4.1.1-5.4.1.3) where $\ell, h, f \in \mathbb{F}$.

The conditions of unity $\Rightarrow \mathbb{F} = \left(\tilde{\mathbb{F}} \cup \tilde{\mathbb{H}} \right) \Rightarrow \mu_{\mathbb{F}}(f) = \left(\mu_{\mathbb{L}}(\ell) \vee \mu_{\mathbb{H}}(h) \right)$ (5.5.1)

The conditions of commonness $\Rightarrow \left(\tilde{\mathbb{F}} \cap \tilde{\mathbb{H}} \right) \Rightarrow \mu_{\mathbb{F}}(f) = \left(\mu_{\mathbb{L}}(\ell) \wedge \mu_{\mathbb{H}}(h) \right)$ (5.5.2)

The conditions of negative relative difference

$$\Rightarrow \tilde{\mathbb{A}} = \left(\tilde{\mathbb{H}} / \tilde{\mathbb{L}} \right) \Rightarrow \mu_{\mathbb{F}}(f) = \left(\mu_{\mathbb{H}}(h) - \mu_{\mathbb{L}}(\ell) \right) \qquad (5.5.3)$$

The conditions of positive relative difference

$$\Rightarrow \tilde{\mathbb{B}} = \left(\tilde{\mathbb{H}} / \tilde{\mathbb{L}} \right) \Rightarrow \mu_{\mathbb{F}}(f) = \left(\mu_{\mathbb{L}}(\ell) - \mu_{\mathbb{H}}(h) \right) \qquad (5.5.4)$$

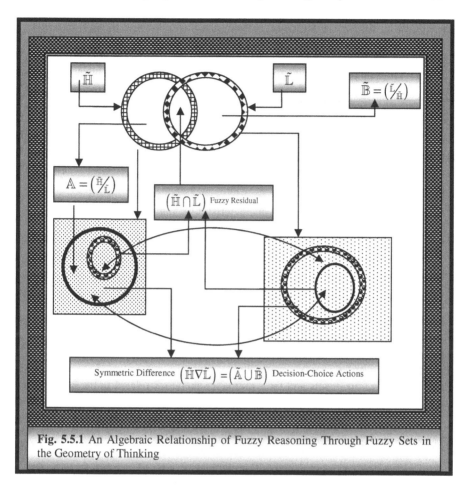

Fig. 5.5.1 An Algebraic Relationship of Fuzzy Reasoning Through Fuzzy Sets in the Geometry of Thinking

The epistemological condition presents a situation where every element in the possibility space is simultaneously knowable and unknowable in the continuum of duality. The conditions of knowability and unknowability may be expressed in terms of epistemic index that combines possibilistic and probabilistic indexes in various weighted proportions. These indexes form the foundation of conditions of acceptability of true-false elements in reasoning by the way of falsification, verification and corroboration as they relate to the exactness and inexactness of scientific propositions.

5.5.1 The True-False Fuzzy Algebra over Inexact Epistemological Space

The fuzzy logic may find expressions in the interactive processes of category, set and group theories under substitution-transformation processes. Let us examine in a little detail the logical structure of the use of fuzzy laws of thought in terms of the classical and fuzzy sets. The basic notion of a set is assumed here in terms of belonging. The essentially analytical components for reasoning must relate to unity, commonness and difference given the identity of the phenomenon of interest. The conceptual *instruments for thought are the relative complement (difference), symmetric difference* and *unity* of characteristic sets. The relative complement relates to the principle of commonness while the symmetric difference relates to the law of unity in terms of becoming and not becoming. In this connection, consider two fuzzy sets $\mathbb{A}$ and $\mathbb{B}$ in terms of general characteristics of an epistemic element with membership characteristic functions $\mu_{\mathbb{A}}(x)$ and $\mu_{\mathbb{B}}(x)$ respectively as opposites in behavior and identity. The unity of the opposites implies that the characteristic set may be written as $(\mathbb{A} \cup \mathbb{B})$ or $(\mu_{\mathbb{A}}(x) \vee \mu_{\mathbb{B}}(x))$, the positive difference implies that we have $(\mathbb{A}/\mathbb{B})$ or $(\mu_{\mathbb{A}}(x) \wedge \mu_{\bar{\mathbb{B}}}(x))$, the negative difference implies that we have $(\mathbb{B}/\mathbb{A})$ or $(\mu_{\bar{\mathbb{A}}}(x) \wedge \mu_{\mathbb{B}}(x))$ and the complete non-interactive opposites may be written in terms of the symmetric difference of $\mathbb{A} \nabla \mathbb{B} = (\mathbb{B}/\mathbb{A}) \cup (\mathbb{A}/\mathbb{B})$ or $\left[(\mu_{\mathbb{A}}(x) \wedge \mu_{\bar{\mathbb{B}}}(x)) \vee (\mu_{\mathbb{B}}(x) \wedge \mu_{\bar{\mathbb{A}}}(x)) \right]$ where $\mathbb{A}$ and $\mathbb{B}$ are the positive and negative characteristic sets respectively.

How are the fuzzy sets and their mutual connections related to the acceptance-rejection decisions of true or false propositions in thought and how may these be related to the conditions of exactness and inexactness? The epistemic decision-choice process is such that we specify the fuzzy sets of false characteristics $\mathbb{F}$ and true characteristics $\mathbb{T}$ with membership functions, $\mu_{\mathbb{F}}(x)$ and $\mu_{\mathbb{T}}(x)$ respectively. The elements in $(\mathbb{F} \cap \mathbb{T}) \Rightarrow (\mu_{\mathbb{F}}(x) \wedge \mu_{\mathbb{T}}(x))$ define the point where the opposites meet and emerge into each other in a continuum to form a unity in an inter-supportive mode. In terms of sciences, this is an exact-inexact continuum where the inexactness fades into the exactness and vice versa. They may be viewed as logical equilibrium points of commonness that will vary from a proposition to a proposition. For a well-behaved membership characteristic function we may write in the true-false duality:

$$\mu_{D}(x) = \begin{cases} \text{Truth if } \mu_{T}(x) \in (\mu_{F}(x^*), 1] \\ \text{False if } \mu_{T}(x) \in [0, \mu_{F}(x^*)) \\ \text{Indifference if } \mu_{T}(x) - \mu_{F}(x) = 0 \end{cases} \qquad (5.5.1.1)$$

The point $x^* \in (\mathbb{T} \cap \mathbb{F})$ is a switch point that imposes a requirement on the decision-choice action as to what will be accepted as true or false. The corresponding membership value defines a point of indifference and is used for decomposing the proposition into true or false statements with their corresponding characteristic sets. The structure of the equation (5.5.1.1) applies also to the exact-inexact duality as to what will be accepted as exactness and inexactness in the knowledge sectors and in the continuum of configuration. The set $(T \cap F)$ may be seen as fuzzy residual in the penumbral region of decision-choice action. Let us note that the condition $\mu_T(x) \in (\mu_F(x^*), 1]$ has its false support with a fuzzy conditionality $\mu_T(x) - \mu_F(x^*) > 0$ and $\mu_F(x) \in [0, \mu_T(x^*)]$ has its truth support with a fuzzy conditionality computed as $\mu_T(x) - \mu_F(x^*) \leq 0$. The case of exactness-inexactness duality, the decision-choice structure, continuum principle and the fuzzy laws of thought are presented in an epistemic geometry in Figure 5.5.2.

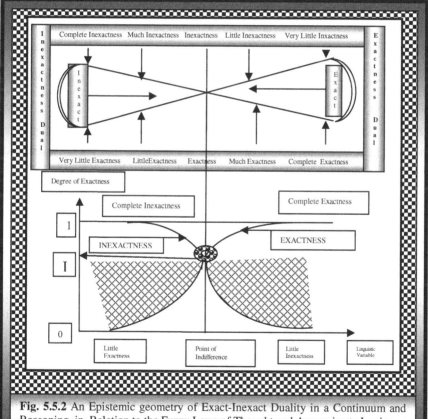

Fig. 5.5.2 An Epistemic geometry of Exact-Inexact Duality in a Continuum and Reasoning in Relation to the Fuzzy Laws of Thought and Approximate Logic

The structure at the level of exact determination and science, the term $\left(\frac{(A\cup B)}{(A\cap B)}\right)$ projects conditions where there is an excluded middle in reasoning such that $(A\cap B)$ is assumed to be a null set or taken to be insignificant in thought, in which case $\left(\frac{A}{B}\right)\cap\left(\frac{B}{A}\right)=\varnothing$ even if $(A\cap B)\neq\varnothing$ which represents the common characteristic set of the excluded middle in the classical reasoning. In this way, $\left(\frac{A}{B}\right)$ has an independent existence and is not connected to $\left(\frac{B}{A}\right)$ which also has its own independent existence and is not connected to $\left(\frac{A}{B}\right)$. The point of discussion applies to either true-false dualism or exact-inexact dualism that is devoid of the common connecting term $(A\cap B)$ by the classical logic of reasoning and the need for simplicity of analysis, representation and decision-choice acceptance. In other words, the complexity is reduced to a simplicity which is attained through the suppression of quality to allow the application of exact rigid determination, hard computing and classical laws of thought.

At the level of inexact flexible determination in knowledge production and science, $\left(\frac{(A\cup B)}{(A\cap B)}\right)$ projects conditions where there is no excluded middle but a continuum in reasoning such that $(A\cap B)$ is not a null set and presents a significance in thought such that $\left(\frac{A}{B}\right)\cap\left(\frac{B}{A}\right)\neq\varnothing$ but is connected by $(A\cap B)\neq\varnothing$ and hence, $\left(\frac{A}{B}\right)$ and $\left(\frac{B}{A}\right)$ reside in independent existence without which their individual identities are not definable and yet are connected by their commonness through a give-and-take relation. Once again the point of epistemic relevance applies to the true-false and exact-inexact dualities that accept the principles of the continuum and contradiction in terms of $(A\cap B)\neq\varnothing$ by the fuzzy logic of reasoning and its laws of thought. The fuzzy reasoning process makes explicit the decision-choice activities of the knower in the knowing process as well as internalizes the knower and his or her subjective characterization of *what the knower knows* and the degree of exactness attached to that which is known in the knowledge production process. In this way, the knower is inside but not outside the social knowledge-production system thus making the knower to interact as well as interfere with the knowledge problem through the membership characterization function of the set, group and category formations. Here, the knowledge-production enterprise is presented as a human creation for discovery through a search from conditions of ignorance to conditions of knowing. It is an inexact and self-correcting system with continual refinements from conditions of inexactness and uncertainty toward conditions of exactness and certainty. Complete exactness and complete certainty are absolutes towards which the activities of the knowledge seeker gravitate; as such, any unqualified claim of a-prior exactness of an area of knowledge production over the epistemological space is an epistemic illusion; to prove its a-priori existence is a phantom problem

[R8.53]. As a rule, the knowledge search process over the epistemological space leads to the discovery of exact-value equivalence and certainty-value equivalence with fuzzy and stochastic conditionalities, but not the absolute.

The fuzzy thought process allows us to examine and understand the epistemic involvement of the knower's coding and decoding through a decision-choice process at all stages in the cutoffs in the final sequence of the epistemic steps. In this setting, an acknowledgement is made of human cognitive limitations composed of linguistic vagueness, quantitative errors, qualitative misunderstandings, institutional constraints and belief systems that impose ideology in the thought process and many others. It is here, that the use of the classical laws of thought encounters important epistemic difficulties, especially in economics, social, biological, medical and other sciences where the subject of the knowledge search is also about the knowledge seeker whose decision-choice actions at every time point complicate the search and the knowing due to continual qualitative changes that create complexity for the knower and his or her relationship with the defective information structure.

The nature of the classical paradigm is to seek universal laws of thought that eliminate subjectivity which is part of the thought process itself. Such elimination implies the elimination of judgment of the knower under interpretive conditions of vagueness and ambiguities that define the penumbral regions on the path of the knower and what the knower believes to know. This externalizes the cognitive agent and imprisons imagination with this structure creating conditions where doubting the familiar and imagining the unfamiliar through thinking outside the classical laws of thought are frowned upon. The result is that all epistemic objects or objects of thought that do not conform to these universal laws are eliminated through the principle of irrationality, paradox or contradiction. The elimination of judgment as part of the knowledge process itself, makes it difficult to apply the classical laws of thought and the methods of exact determination to the study and understanding of self-exited, and self-corrected and self-organizing systems that are dynamic and holistic objects of the laws of thought, of course, including the knowledge-production system as a self-correcting system. The study of these self-exiting, self-organizing and self-correcting systems, such as social systems, biological systems, climatological systems, environmental systems, medical systems, that require methods of organicity, systemicity, informatics, synergetics and complexity theory, presents challenges to cognition in the modern direction of sciences where interdisciplinary approaches impose special accounts of simultaneous interactions of quality, quantity and time.

Chapter 6
Zones of Thought:
Reflections on the Theories of Thought

Let us reflect on the zones of thought and how they are related to the fuzzy and classical paradigms and laws of thought in cognition as the cognitive agents journey on the epistemological highway of knowing to the house of knowledge. The epistemological space is defined by the type of information structure assumed which in turn defines the type of the set of laws of thought needed for its processing. The relevant variables characterizing the information contents for the epistemic processing will be made explicit and the conditions for representational exactness and inexactness will be stated for true-false acceptance in the relevant zones of thought. The knowledge content of the processed information in each zone will be examined and related to the epistemic conditionality. In the previous chapter, we examined exactness of inexact science and how it relates to the concepts of opposite, duality, continuum and the justification for the use of the fuzzy paradigm with its laws of thought and mathematics. It may be noticed that the fuzzy laws of thought satisfy its internal requirement for true-false acceptance in all conditions of the knowledge production in that every epistemic element contains true-false characteristics including the statement of the fuzzy laws of thought. The classical laws of thought cannot satisfy its condition of true-false evaluation as observed by Russell which leads to many philosophical discussions on the *principle of excluded middle*.

6.1 Zones of Unity of Exact and Inexact Sciences

The knowledge-production process of *inexact sciences* is not different from that of the *exact science*. In other words, there is no exact or inexact science but science where every knowledge area simultaneously contains exactness and inexactness in unity that appear in relative degrees whose separation is under the principle of fuzzy conditionality. Both the exact and inexact sciences follow the general epistemic path of the *knowledge square* proceeding from the space of cognitive potential and moving through the possibility and probability spaces to the space of cognitive actual where the knowledge content claimed is always under a form of epistemic conditionality. This is the universal principle of the knowledge-production process embodied in the theory of the knowledge square as presented

K.K. Dompere: Fuzziness and Found. of Exact and Inexact Sci., STUDFUZZ 290, pp. 103–123.
springerlink.com   © Springer-Verlag Berlin Heidelberg 2013

in [R2.10]. In the epistemic process of knowing over the epistemological space, the cognitive agents work within the conditions of simplicity-complexity duality. The truths of the objects of thought are distributed in a continuum of extreme simplicity to an extreme complexity that defines the simplicity-complexity duality as opposite poles which are mutually connected by a give-and-take relation. Simplicity and complexity in the objects of thought relate to static-dynamic and quality-quantity dualities with time functioning in the conditions of neutrality.

Over the epistemological space, the objects under the laws of thought acquire attributes of simplicity if they are qualitatively static (for example a particle under quantitative motion) and further simplicity if they are quantitatively static (for example a particle at rest or a simple mathematical object). The objects of thought acquire attributes of increasing complexity if they are both quantitatively and qualitatively dynamic in terms of simultaneity of complex motions of quantity and quality. The quantitative motion is seen in terms of time-space phenomenon where the measure of quantity may be specified as a mathematical variable or as a linguistic variable. The qualitative motions defined in terms of qualitative transformations or categorial conversions are seen in terms of time-quantum phenomenon that must be analytically acknowledged with the passage of time. The simultaneous existence of quality-quantity motions produces extreme complexity for cognition (for example societies as self-organizing and self-correcting systems, or a living entity or a biological object that is also a self-exiting and self-correcting system). It is here that the study of socio-economic development, medical sciences, dark-matter-energy phenomenon, and phenomenon of the behavior of the universe acquires increasing complexity and increasing analytical difficulties when one works with the classical paradigm of thought where the theory of truth does not admit its opposite of falsehood. Dealing with this exact-inexact problem or true-false problem or the dilemma of the classical paradigm with its law of thought, Russell states the following:

1. *Our theory of truth must be such as to admit of its opposite, falsehood.*
2. *It seems fairly evident that if there were no belief there could be no falsehood, and no truth either, in the sense in which truth is correlative to falsehood.*
3. *...it is to be observed that the truth or falsehood of a belief always depends upon something which lies outside the belief.*

   In accordance with our three requisites, we have to seek a theory of truth which (1) allows truth to have an opposite, namely falsehood, (2) makes truth a property of beliefs, but (3) makes it a property wholly dependent upon the beliefs of outside things [R14.100, p.120-123].

The relevance of these statements of Russell must be seen in the context of true-false duality with a continuum where the fuzzy theory of truth meets all the requisites stated above as well as applicable to exact-inexact phenomena. Within the duality, the opposites exist in some form of correspondence between information and beliefs where the correspondences are established by cognitive transformation functions that establish the information processing activities over the epistemological space as illustrated by the geometry of the theory of the knowledge square. Generally, the information over the epistemic space is defective and hence by extension, the belief system is defective that deprives any cognitive agent epistemic absolutism of exactness and truth that are abstracted epistemic activities.

The cognition functions under defective information structure gives rise to exact-inexact and certainty-uncertainty dualities where there is no either complete exactness or complete inexactness but simultaneous approximation through judicious judgment in relation to simplicities and complexities. A complete simplicity is when the objects under the laws of thought are both qualitatively and quantitatively static while a complete complexity is encountered when the epistemic objects are qualitatively and quantitatively dynamic to give rise to four zones of thought. These relations are presented in Figure 6.1.1 showing the four zones of the knowledge search with varying complexities. The four zones are defined by the interactions of simplicity-complexity duality and statics-dynamics duality. The four zones are defined in the epistemological space but not in the ontological space. By zonal analysis we can abstract the conditions of paradigm of thought, truth and the supporting epistemic conditionality. Each one of the zones has a corresponding information structure. These information structures may be viewed as categories of information structure one of which may be taken as the primary category from which others emerge as logical derivatives.

In the complete complexity, cognitive agents are challenged with the simultaneous understanding of the behaviors of both qualitative and quantitative motions in which case we must deal with the complex phenomena in the three dimensional space of quality, quantity and time. The epistemic ingenuity and creative imagination in representation and cognition are reflected in how well we are able to deal with this quality-quantity simultaneity as time alters. The quality-quantity duality in interactive models of statics-dynamics duality within the substitution-transformation processes spins the categories of logical areas of thought and the knowledge sectors. The conditions of these sectors point to the direction to choose the appropriate laws of thought. In the previous times of the knowledge-production process, we have only available to us the classical paradigm with its laws of thought and corresponding mathematics. The uses of the classical paradigm require the imposition of assumptions to restrict the problem into the zone of logical convenience. These required assumptions of technical convenience in most cases render the tools of the knowledge finding irrelevant to the needed knowledge in some areas of cognition. This is particularly so in social sciences, medical sciences, biological sciences, complexity sciences and many more where quality and qualitative motion most often overrides conditions of quantitative dispositions. This is a particular problem in the theoretical study of social change or socioeconomic development, developmental biology, psychological sciences and others.

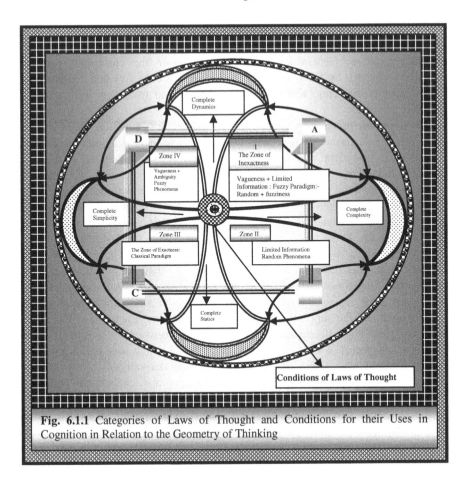

Fig. 6.1.1 Categories of Laws of Thought and Conditions for their Uses in Cognition in Relation to the Geometry of Thinking

To see the logical difficulties in operating in the quality-quantity-time space, let us examine the zonal structures in Figure 6.1.1. There are four categorial zones that require attention and how they relate to exact and inexact sciences and then to the applicable areas of laws of thought of the classical and fuzzy paradigms. From Figure 6.1.1, Zone I defines a complete inexactness that corresponds to the presence of *fuzzy-stochastic uncertainty* and *stochastic-fuzzy uncertainty* in the application of the laws of thought. Here, the objects of the laws of thought are dynamic and organic with substantial defective information structure for processing. It is under this zone that complexities arise and present challenges to the use of the classical paradigm and its laws of thought to process information on quality, quantity and time for the knowledge production. The mathematical space corresponding to this zone is fuzzy-stochastic or stochastic-fuzzy space with fuzzy-random or random-fuzzy variables respectively that must be symbolically represented and manipulated by appropriate logical or mathematical operators. The *fuzzy-random variable* or the *random fuzzy-variable* generates vague or inexact probabilities [R3.53] [R4.48] [R4.57][R4.79][R4.91]. It is within this zone that the fuzzy paradigm presents an important approach to reasoning and knowledge production where the fuzzy laws of thought allow reasonable representation of the information-knowledge structures that account for relevance and rigor within complexities under defective information structures.

In this zone, the symbolic representation must account for fuzziness (quality, vagueness, and ambiguity) in the thought process and appears as three supporting elements of magnitude (value) and the *degree of exactness* (possibility measure or fuzzy characteristic function) that lies between zero and one with *degree of certainty* (probability measure) that lie between zero and one and their interactive value that also lies between zero and one where the interactive value (degree of exact or inexact probability) may be either separable or non-separable. Our current inexact-exact duality in classification of science places inexact science in Zone I. The conditions of human cognition, however, seem to suggest that all potential knowledge objects from the *exact ontological space* belong to Zone I with varying degrees of epistemic complexities depending on the manner in which the information structure is constructed. The analytical point of emphasis is that exactness in any area of the knowledge production, hence exact science is claimed by decision-choice action with fuzzy-stochastic conditionality, and objectivity is claimed by assumption. Zone I is the primary category from which other epistemological descriptive categories tend to emerge. The claim of unconditional exactness in this zone for any area of knowledge production over the epistemological space is not different from the claim of absolute truth in the epistemological space. What we can claim is that the exact knowledge and exact science ontologically exist and our epistemic process is continually in fuzzy-stochastic tuning modes towards the ultimate. The characteristics of this zone may be summarized as a) inexact possibility, b) inexact and incomplete information (inexact probability), c) fuzzy-random or random-fuzzy variable for information representation, d) fuzzy-stochastic or stochastic-fuzzy conditionality for claims of knowledge items and d) fuzzy paradigm.

Zone III presents an artificially created information-knowledge sector to which exact sciences belong. It is an inexact epistemic representation of exact ontological items for cognition due to assumptions imposed on the epistemological space and the nature of cognitive limitations in the process of knowing. The zone is a logical derivative of the zone I and corresponds to the presence of exact certainty and logic in the knowledge production process where variable representations are exact, non-random and non-fuzzy with exact rigid determination under the classical paradigm. The zone contains non-defective information structure in the sense of being complete and non-vague. The mathematical space for the analysis and synthesis of knowledge problems is non-stochastic and non-fuzzy topological space where exact classical variables are used in the information-knowledge representation, and can be manipulated with the classical laws of thought that can be applied in any area of knowledge production whose structural characteristics can meet these conditions of the information structure. The non-random and non-fuzzy variables are created by implicit and explicit assumptions to simplify complexities. Here, both quality and subjectivity are frozen throughout the investigation and knowledge production in order to concentrate on the exact *quantity-time* relations that correspond to *basic forms* of thinking. The objective is to allow the direct and indirect applications of the classical laws of thought which further eliminates or simplifies the role of subjective judgment in the decision-choice action for the knowledge production. The epistemological space of the knowledge search is simple and exact which allows the construction of the knowledge house of truth without its opposite of falsehood. Here, exactness and objectivity may be claimed for this sector of the knowledge production but not as a-prior conditions. In this space, the pillars of the three-dimensional relations of quality, quantity and time are collapsed into quantity-time relation that connects to the space-time phenomenon. The space, therefore, is non-holistic and its logic is applicable to substantially restricted objects that are suitable for the use of the classical laws of thought where exactness and truth can be claimed without falsehood.

It must be brought to focus that for the objects of thought, in this space, that correspond to Zone III, to meet the requirements of the use of the classical paradigm and laws of thought, the following conditions must be met.

1. Acceptance of identity as existing in dualism with clearly defined non-interactive opposites in a contradictory mode;
2. Elimination of contradiction in the identity due to quality;
3. Elimination of contradiction through constantly dividing the whole into smaller and smaller parts of micro-units until the quality is constant and the contradiction is no longer present in the identity;
4. The use of analytical methods (from micro-to-macro) in the single units that may be referred to as micro-analytics;
5. The application of the classical laws of thought encounters difficulties and sometimes fails when one leaves the space of methodological analytics and enters into the space of methodological synthesis for the whole or the organicity or the systemicity in the presence of relational continuum in both qualitative and quantitative spaces.

These requirements of the classical laws of thought are applicable in the behavior of space of quantity-time phenomena. They, thus, exclude any meaningful and relevant discussions on continuum phenomenon in the quality-time processes. At the level of motion, the continuum hypothesis is applied only to space time phenomena in a manner that has nothing to relate to the concept of quality-quantity duality where human thought is challenged with judicious application of judgment that rests on subjectivity but not on objectivity. In this zone, the epistemic conditionality vanishes and information becomes knowledge since the epistemological space is in isomorphic relation with the ontological space in terms of exactness and completeness regarding the phenomenon of concern. There is no unsureness of this information-knowledge structure as an input into other decision-choice activities. The characteristics of this zone may be summarized as a) exact possibility, b) exact and complete information structure (no probability), c) exact epistemological space d) non-fuzzy and non-random (classical exact) variable for information representation, e) the use of classical paradigm of thought and f) no epistemic conditionality for claims of knowledge items.

Zone II deals with decreasing complexity with quality held relatively constant and where an increasing analytical complexity is the result of limited or incomplete information which may further be constrained by some vagueness and ambiguities in decreasing magnitudes. The zone contains defective information structure in the sense of being incomplete and non-vague. The Zone is a logical derivative of Zone I as the primary category where the fuzziness is stripped off to retain the conditions of exactness. The mathematical space corresponding to this zone is non-fuzzy and stochastic topological space where outcomes are uncertain but with exact values in the outcomes. The information representation takes the form of *exact random variables,* and the corresponding exact probability distributions where fuzziness is taken to be small and suppressed by assumption. Here, vague or inexact probabilities are discounted in a manner where the knowledge production under ambiguities is logically non-existent. This allows the application of the principle of exact symbolism and for the classical laws of thought to be applied to the epistemic objects in Zone I since quality is held constant and unchanging with vagueness and ambiguity assumed away through the exact symbolic representation. The only relevant constraint on thought is quantity of information that defines the uncertainty of knowledge as decision-choice input. Additionally, the incomplete information available for epistemic processing is assumed to be exact with exact probability values that are used to specify the degree of uncertainty associated with claims of knowledge. The symbolic representation in the thought process appears as events in two supporting elements of magnitude (value) and degree of certainty (probability value) where the degree of exactness is always equal to one, the degree of certainty lies between zero and one which constitutes the stochastic conditionality and where objectivity is assumed to be present and subjective judgment is assumed away. The admittance of subjectivity in this zone moves us into personal probability space. The characteristics of this zone may be summarized as a) exact possibility, b) exact and incomplete information structure (exact probability), c) exact epistemological space d) non-fuzzy and random (classical exact random) variable for information representation, e) the use of classical paradigm of thought and f) stochastic conditionality for claims of knowledge items.

The epistemic Zone IV defines an information-knowledge area of increasing complexities where vagueness and information limitation are increasing with increasing qualitative dispositions and values. It is also a logical derivative from the zone I as the primary logical category in the epistemological space. In this epistemic zone, quality and subjective judgment acquire an increasing importance in the decisions on true-false propositions and exact-inexact information-knowledge representation and claims. This is fuzzy and complete information epistemological space. The mathematical space that corresponds to this area of knowledge is non-stochastic and fuzzy topological space with *non-random fuzzy variable* and corresponding possibility distribution. In this space, information is assumed to be inexact but full with small probability that may be neglected by assumption. Quality and linguistic variables take the central stage of the knowledge search with internalization of the knowledge seeker. The symbolic representation in the thought process appears as two supporting elements of magnitude (value) and the *degree of exactness* that lies between zero and one with certainty equals to one, and where the degree of exactness specifies the condition to which the magnitude belongs to the set of exactness. The information-knowledge process follows the fuzzy paradigm with the principle of continuum and the fuzzy laws of thought. The zone is characterized by fuzzy uncertainty which is sometimes misunderstood by knowledge seekers for probabilistic uncertainty and analyzed with the use of toolbox of the classical paradigm and probability. The characteristics of this Zone IV may be summarized as a) inexact possibility, b) inexact and complete information structure (no probability), c) inexact epistemological space d) fuzzy and non-random variable for information representation, e) the use of fuzzy paradigm of thought and f) fuzzy conditionality for claims of knowledge items.

The classical paradigm with its laws of thought does not provide us with channels for dealing with quality and subjective judgment and hence their application either fails or comes to confront some logical difficulties. The results of the analyses are the presence of paradoxes and contradictions in the applications of the laws of thought within the classical paradigm (see Arrows possibility theorem). In Zone IV, the fuzzy laws of thought may be brought into action to deal with the conditions of fuzzy uncertainty, quality, linguistic quantities and the need for subjective action. Both Zones II and IV represent partial uncertainty of the epistemic process in a manner where exactness and certainty equivalences are conditional claims and where some epistemic items fall outside the classical laws of thought.

In our current knowledge production, the exact sciences, as they are claimed, belong to Zones II and III where the classical paradigm and its laws of thought with excluded middle are applicable since the information structure is either exact and complete or exact and incomplete. The classical paradigm has nothing to say about the behavior of epistemic elements in Zones I and IV except when their attributes are reduced to conform to those in Zones II and III by assumption. This is the case, for example, with the studies in social, medical and biological sciences and with increasing difficulties encountered with the studies in complexity science, organicity, systemicity, synergetics and energetics. The inexact sciences, as seen from the classical ideology of knowledge production, fall into Zones I and IV. The fuzzy paradigm with its laws of thought, however, is applicable to all the information-knowledge zones of, I, II, III and IV. The discussions in this monograph suggest that there is no area of knowledge that is free from inexactness and uncertainty. The general applicability of the fuzzy paradigm with its laws of thought to all propositions leads to a statement of *fuzzy completeness theorem* which is derived from the principle that the epistemological space in which cognitive agents work for different knowledge sectors is defined by defective information structure composed of vagueness and incompleteness. In the social sciences, the uncertainties of the defective information structure may be enhanced by deceptive information structure composed of disinformation and misinformation characteristics.

Theorem 6.1.1: Fuzzy Completeness Theorem

Every true classical proposition has a fuzzy covering in a manner where truths and falsities exist in continua and in an inseparable unity of thought in such a way that every classical topology has a fuzzy topological covering but the converse is not true. Every exact symbolism has corresponding vague symbolism as a fuzzy covering.

NOTE:

The core of the fuzzy completeness theorem must be viewed in terms of concepts of category, set and group systems. In terms of set, it is simply $\{0,1\} \subset (0,1)$ in terms of degrees of belonging.

Theorem: 6.1.2: Fuzzy Inexactness Theorem

As an epistemic object increases in complexity, an exact symbolic representation of the thought of its behavior and useful understanding of such behavior are both inconsistent and incomplete at the threshold of cognition, and hence inexact in the knowledge production in such a way that all claimed exact conclusions are partial and derived from imprecise and vague premises with subjective judgments through decision-choice actions in the quality-quantity-time space.

NOTE:

The fuzzy inexactness theorem may be stated symbolically by introducing complexity index, β into the membership characteristic functions of the positive and negative duals of $\mu_{\mathbb{X}^P}(x\,|\,\beta)$ and $\mu_{\mathbb{X}^N}(x\,|\,\beta)$ respectively. In this way, β becomes a parametric shifter that moves both the positive and negative functions bodily to the right as complexity increases and to the left as complexity decreases. The body shifts are not necessarily equal. The outcomes of the positive-negative characteristic combinations as the complexity increases or decreases will depend on the elasticity of the complexity of relative changes of negative and positive qualitative dispositions of the system. The elasticity of the complexity index indicates the responsiveness of the negative-positive qualitative changes to the system's transformations as viewed in the give-and-take relationality.

The two theorems guarantee a system of exact reasoning with fuzzy covering as well as sequential in exact reasoning (approximate reasoning) in the fuzzy domain that allow propositions to be generated and examined in the four epistemological zones of I, II, III and IV to account for continuity of thought from simplicity to complexity, and from complete inexactness to complete exactness through human decision-choice actions with mathematical and logical rigor. The fuzzy mathematics may be seen as mathematical study of the phenomenon of vagueness and ambiguity. The strength of the fuzzy paradigm with its laws of thought lies in the idea that it is applicable to qualitative and quantitative characteristics with time operating as a neutral element and where most qualitative characteristics are quantified by linguistic numbers. The fuzzy paradigm allows us to use mathematical reasoning in vague and ambiguous spaces and to compute approximations from imperfections to perfections and from inexactness to exactness as cognitive characteristics of human behavior, where error-correction process is part of the human actions in all endeavors of life. Such error-correction process is called fuzzy tuning that is consistent with human experience of knowledge production as self-contained and self-correcting system.

This is consistent with the notion that the knowledge structure, at any time point, is limited in scope and inexact by the previous information-knowledge structure, characteristics of current experience and available technology and methods of knowing. In this way, the information-knowledge structure in support of the decision-choice actions is defective in terms of completeness, exactness, assumptions, ideas, conclusions and certainty, and where all of these exist in dualities with continuum. As such, the continuum phenomenon in which truth, falsity, exactness and reality are defined in stage-by-stage substitution-transformation processes, in a hierarchical order of refinement, affirms the validity of fuzzy logical thinking and thought in exactness that supports inexact sciences and exact sciences. Thus, exact science is a derivative from the **inexact** science by categorial conversion where inexactness is the primary category of human thought. The fuzzy paradigm and the corresponding laws of thought allow us to derive truth values in exact and certainty equivalences with epistemic conditionality. This is how our knowledge production proceeds; and this is how the exact and inexact sciences must be viewed and specified with epistemic conditionality.

6.2 The Methods and Techniques of Reasoning in the Fuzzy Space

To understand the method of reasoning in the fuzzy space with the fuzzy laws of thought we may specify the conceptual toolbox. Let us consider a general set of epistemic objects, Ω with generic element $\omega \in \Omega$ which we shall called the epistemic set. The set structure of the Ω will vary over different knowledge areas. The toolbox for the study of each $\omega \in \Omega$ contains set, category, opposites, polarity, duality, and continuum. Every epistemic object is described and distinguished by its characteristic set $\mathbb{X}$ which places it in a category with a membership characteristic function of the form $\mu_{\mathbb{X}}(x) \in [0,1]$. The characteristic set is structured in terms of opposites, polarity, and duality with relational continuum. The defining characteristic set is specified by two characteristic sub-sets of negative disposition $\mathbb{X}^{N}$ and positive dispositions $\mathbb{X}^{P}$ with $\mathbb{X} = \left(\mathbb{X}^{P} \bigcup \mathbb{X}^{N} \right)$ that defines the identity condition. The existence of the positive and negative characteristic sets allows the specification of duality and polarity for every epistemic problem. The sorting process of the characteristics into negative and positive for any $\omega \in \Omega$ is accomplished by a choice of membership characteristic functions of $\mu_{\mathbb{X}^P}(x)$ and $\mu_{\mathbb{X}^N}(x)$ for the positive and negative sets respectively, such that $\mu_{\mathbb{X}^P}(x) = \left(\mu_{\mathbb{X}^P}(x) \vee \mu_{\mathbb{X}^N}(x)\right)$. The relational fuzzy cardinality, (Fcard.) of the sets may be specified as:

$$\int_0^\infty \mu_{\mathbb{X}}(x)dx = \int_0^\infty \mu_{\mathbb{X}^P}(x)dx = \int_0^\infty \mu_{\mathbb{X}}(x)dx \overset{=}{=} \text{Fcard}\mathbb{X}=\text{Fcard}\mathbb{X}^P = \text{Fcard}\mathbb{X}^N \quad (6.2.1)$$

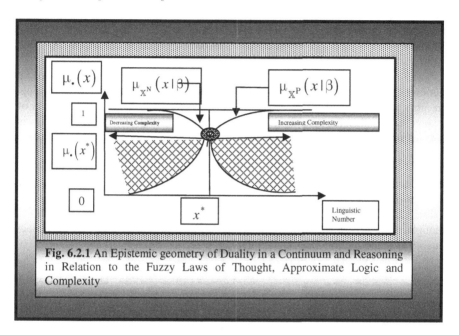

Fig. 6.2.1 An Epistemic geometry of Duality in a Continuum and Reasoning in Relation to the Fuzzy Laws of Thought, Approximate Logic and Complexity

The problem of true-false conclusions about each $\omega \in \Omega$ is formulated as a reconciliation of the negative and positive characteristic sets from the available information structure. The negative and positive characteristic sets may have different interpretations for different problems. For example, the concept of inexactness may be represented as the negative characteristic set with a corresponding distribution of the degrees of belonging $\mu_{\mathbb{X}^P}(x)$ while the concept of exactness may be represented as the positive characteristic set with corresponding distribution of degrees of belonging $,\mu_{\mathbb{X}^P}(x)$. The functional structures of the positive and negative characteristic sets are such that the positive characteristic set is increasing in $x \in \mathbb{X}$ and negative characteristic set is decreasing in $x \in \mathbb{X}$ as they are shown in Figure 6.2.1. We specify negative-positive duality in relation to the study and analysis of the phenomenon of national interest. The analytical concept of duality and the principles of continuum and contradiction are used in specifying the poles and the polarity. In this way, there is a rejection of the Aristotelian basic principle of excluded middle in the classical laws of thought, the uses of which give rise to paradoxes in classical mathematics, theoretical sciences and axiomatic systems such as social choice theory, mathematical physics, mathematical economics and other. The identity of an object $\omega \in \Omega$ is established by the relational structure of negative-positive characteristics in terms of the strength of belonging. The negative and positive characteristics exist in a complex continuum with logical give-and-take relationship that helps to define their mutual existence, negation and the fuzzy residuals in concepts and thoughts. The separation of the negative and positive characteristics is a decision-choice problem requiring information and computational schemes and subjective analysis of quality.

To ascertain the information on negative and positive characteristics, we use various instruments of acquaintances to build the relevant database for processing. In other words, two information bits are constructed with various instruments of acquaintances that may be done separately. The rationale for such approach is derived from possible over evaluations if benefits are anticipate for either negative or positive characteristics and under evaluations if costs are anticipated for either negative or positive characteristics since cost-benefit duality exists in decision-choice tension that provides the force of resolution between negative and positive characteristics in establishing the identities. Here, we should not lose sight of axiomatic and empirical conditions in establishing the information basis of thought. The structure of negative-positive duality representing inexact-exact duality as a continuum phenomenon may be stated as Figure 6.2.2.

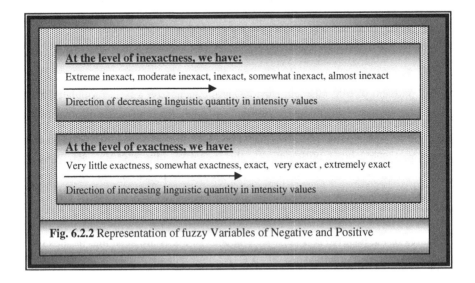

At the level of inexactness, we have:

Extreme inexact, moderate inexact, inexact, somewhat inexact, almost inexact

Direction of decreasing linguistic quantity in intensity values

At the level of exactness, we have:

Very little exactness, somewhat exactness, exact, very exact, extremely exact

Direction of increasing linguistic quantity in intensity values

Fig. 6.2.2 Representation of fuzzy Variables of Negative and Positive

The exact pole is one whose exact characteristic set outweighs the inexact characteristic set. Similarly, the inexact pole is one whose inexact characteristic set outweighs the exact characteristic set. The point of inexact-exact indifference is when the two sets are perceived to be equal in terms of the degrees of belonging. This is amplified in Figure 6.2.3 where exactness is represented by a positive characteristic set and inexactness is represented by a negative characteristic set.

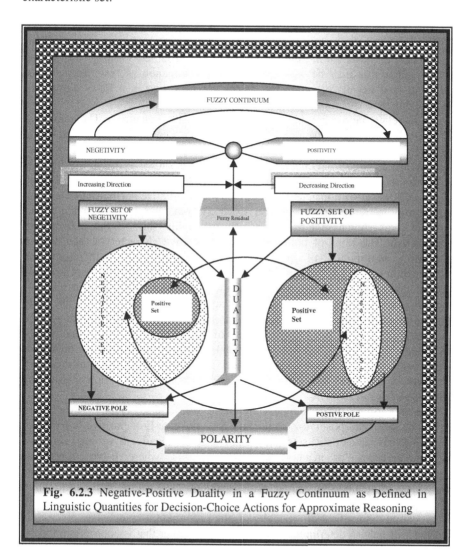

Fig. 6.2.3 Negative-Positive Duality in a Fuzzy Continuum as Defined in Linguistic Quantities for Decision-Choice Actions for Approximate Reasoning

When the two sets are joined together we have a structure of continuum phenomenon as in Figure 6.2.3 that presents a situation where there are two sets of characteristics in any of the objectives in terms of its negative or positive in supporting the phenomenon of interest. It is analytically useful to view them in the decision-choice process in terms of interactions between polarities and dualities under soft or fuzzy computing. The concepts of positive and negative are linguistic variables that exist in the space of vagueness, ambiguity and subjectivity and hence are fuzzy notions that may be represented as fuzzy variables. The negative-positive duality provides us with two opposite fuzzy variables whose degrees of belonging exist in a continuum of linguistic quantities for decision-choice actions as shown in Figure 6.2.1. As fuzzy variables, we may consider the set of values that may be assigned to each phenomenon as two fuzzy sets of positivity and negativity where the values are measured in a fuzzy domain to include linguistic quantities.

Let us relate the discussions on the negative-positive duality to Theorem 6.1.2 of Fuzzy Inexactness Theorem in terms of increasing complexity and exact symbolism for reasoning. First, an increase in complexity is shown in terms of an increase in the size of the characteristic sets. This is shown as a rightward bodily shift of the membership characteristic functions of the negative and positive characteristic sets with increasing multiplicity of qualitative attributes that make it less useful to represent it with an exact symbol. Similarly, a reduction in complexity is shown in terms of a leftward bodily shift of the membership characteristic functions of both negative and positive characteristic sets with a decreasing multiplicity of qualitative attributes. In general, therefore, the specification of the membership characteristic functions must include complexity parameter which must be the same for the membership characteristic functions for the negative and positive characteristic sets. This is discussed in the note immediately after the theorem. The choice of the complexity parameter will be dictated among other things by the nature of the phenomenon. The principles of the category, opposites, duality, polarity and continuum impose on the analytical structure the need to choose an appropriate functional representation of the fuzzy numbers for reasoning under fuzzy paradigm and corresponding fuzzy rationality with fuzzy laws of thought.

The appropriate fuzzy numbers are those whose membership characteristic functions are of the positive-Z or positive-S or positive-E or positive-R types for the positive characteristic sets and negative-Z, or negative-S or negative-E or negative-R types for negative characteristic sets, where E-type represents exponential class of membership characteristic functions and R-type represents ramp class of membership characteristic functions. The nature of the information signals regarding the negative and positive characteristic sets on the basis of which fuzzification-defuzzification module of the information-knowledge process is created will help to determine the choice fuzzy number that is appropriate for the epistemic object of fuzzification. Examples of the classes of these fuzzy numbers for characterizing negative-positive duality in continuum that meets the fuzzy laws of thought are provided in section 6.3.

6.3 Relevant Fuzzy Numbers for Fuzzy Reasoning and Computing in Opposites, Duality, Polarity Continuum, Category and Unity

Consider the reference set of non-negative Ω of characteristic set. Two classes of fuzzy sets with corresponding membership functions describing the identity of any epistemic element $\omega \in \Omega$ may be examined: a) "x is a negative characteristic", b) "x is a positive characteristic" where $x \in R^+$ is an evaluated linguistic value from the researcher. Let $\mu_.(x) \in M$ be a generic membership function defined over M with $\mu(x) \in [0,1]$, and where M is a set of membership functions for both negative and positive sets. We shall first deal with membership functions that express the concept of "x is negativity" as an evaluation for $\omega \in \Omega$. The negative-positive duality is general that captures cost-benefit duality, false-true duality, injustice-justice and may be replaced by any duality of epistemic relevance.

6.3.1 Fuzzy Numbers for "x Is Negative (inexact)"

Let us examine a class of fuzzy numbers that relate to the statement of "x is irrelevant" as an evaluation of $\omega \in \Omega$. These statements are considered as linguistic variables.

6.3.1.1 An Exponential Fuzzy Number for "x Is Negative" Description for $\omega \in \Omega$

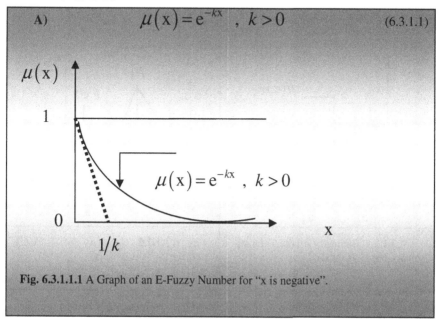

$$\mu(x) = e^{-kx} \quad , \quad k > 0 \tag{6.3.1.1}$$

Fig. 6.3.1.1.1 A Graph of an E-Fuzzy Number for "x is negative".

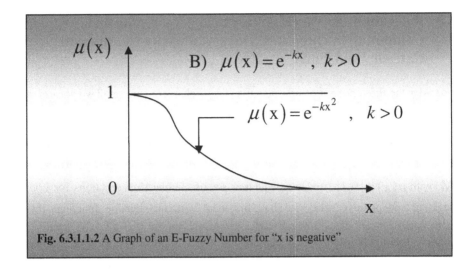

Fig. 6.3.1.1.2 A Graph of an E-Fuzzy Number for "x is negative"

6.3.1.2   The Z-Fuzzy and R-Fuzzy Numbers for "x Is Negative"

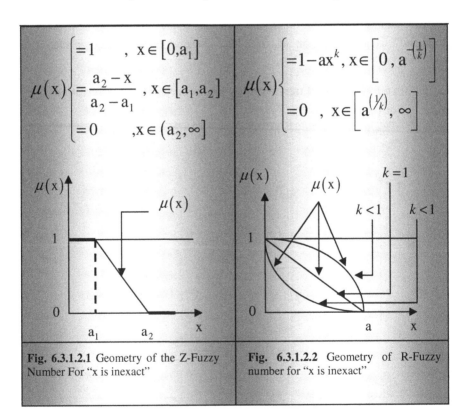

Fig. 6.3.1.2.1 Geometry of the Z-Fuzzy Number For "x is inexact"

Fig. 6.3.1.2.2 Geometry of R-Fuzzy number for "x is inexact"

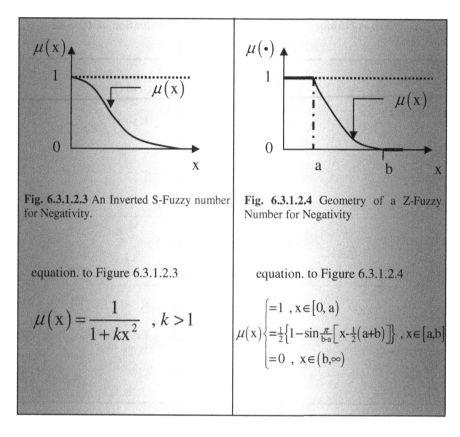

Fig. 6.3.1.2.3 An Inverted S-Fuzzy number for Negativity.

Fig. 6.3.1.2.4 Geometry of a Z-Fuzzy Number for Negativity

equation. to Figure 6.3.1.2.3

$$\mu(x) = \frac{1}{1+kx^2} \quad , k > 1$$

equation. to Figure 6.3.1.2.4

$$\mu(x)\begin{cases} =1 \ , x\in[0,a) \\ =\frac{1}{2}\left\{1-\sin\frac{\pi}{b\text{-}a}\left[x\text{-}\frac{1}{2}(a\text{+}b)\right]\right\} \ , x\in[a,b] \\ =0 \ , x\in(b,\infty) \end{cases}$$

6.3.2 Fuzzy Nmbers for "x Is Positive (Exact)" as an Evaluation of $\omega \in \Omega$

Let us now turn our attention to fuzzy numbers that tend to express the linguistically numerical idea of positive, such as "a is positive". Just as the linguistic variable "x is negative", we can express positive in varying degrees in a continuum.

6.3.2.1 E-Fuzzy Number for "x Is positive" as an Evaluation of $\omega \in \Omega$

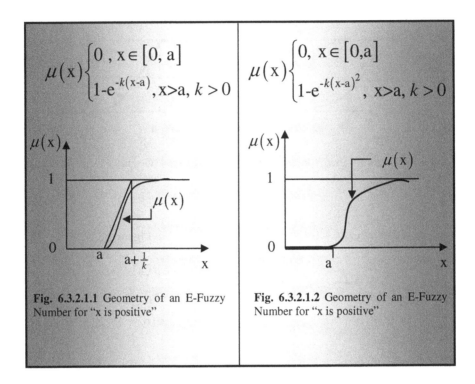

$$\mu(x)\begin{cases}0\,,\,x\in[0,\,a]\\1\text{-}e^{-k(x\text{-}a)}\,,x>a,\,k>0\end{cases}$$

$$\mu(x)\begin{cases}0,\,x\in[0,a]\\1\text{-}e^{-k(x\text{-}a)^2},\,x>a,\,k>0\end{cases}$$

Fig. 6.3.2.1.1 Geometry of an E-Fuzzy Number for "x is positive"

Fig. 6.3.2.1.2 Geometry of an E-Fuzzy Number for "x is positive"

6.3.2.2 S- and Ramp Fuzzy Numbers for "x Is Positive"

Let us now give examples of functional structure S-type and R-type of Fuzzy numbers. Keep in mind that these fuzzy numbers can be approximated by either TFN (triangular fuzzy number) or TZFN (trapezoidal fuzzy number) fuzzy numbers depending on the nature of the opposites, polarity, duality and corresponding continuum. The functional examples, geometric examples and corresponding equations are given in Figures (6.3.2.2.1-6.3.2.2.4).

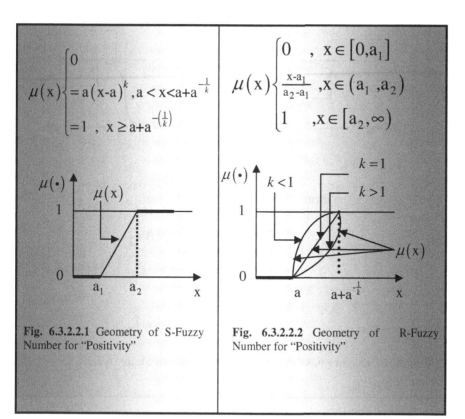

$$\mu(x)\begin{cases}0\\=a(x\text{-}a)^{k},a<x<a+a^{-\frac{1}{k}}\\=1\;,\;x\geq a+a^{-\left(\frac{1}{k}\right)}\end{cases}$$

$$\mu(x)\begin{cases}0 & ,\;x\in[0,a_{1}]\\\frac{x\text{-}a_{1}}{a_{2}\text{-}a_{1}} & ,x\in(a_{1},a_{2})\\1 & ,x\in[a_{2},\infty)\end{cases}$$

Fig. 6.3.2.2.1 Geometry of S-Fuzzy Number for "Positivity"

Fig. 6.3.2.2.2 Geometry of R-Fuzzy Number for "Positivity"

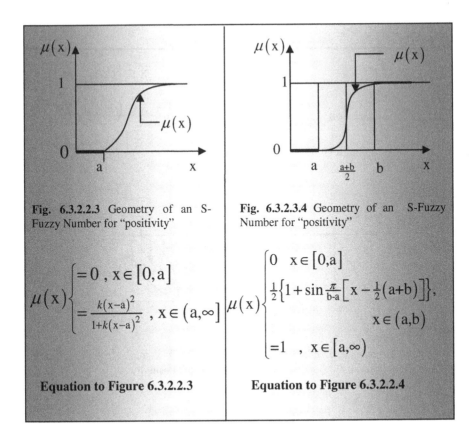

Fig. 6.3.2.2.3 Geometry of an S-Fuzzy Number for "positivity"

Fig. 6.3.2.3.4 Geometry of an S-Fuzzy Number for "positivity"

$$\mu(x)\begin{cases} =0 \ , \ x \in [0,a] \\ =\dfrac{k(x-a)^2}{1+k(x-a)^2} \ , \ x \in (a,\infty] \end{cases}$$

$$\mu(x)\begin{cases} 0 \quad x \in [0,a] \\ \frac{1}{2}\left\{1+\sin\frac{\pi}{b-a}\left[x-\frac{1}{2}(a+b)\right]\right\}, \\ \qquad\qquad x \in (a,b) \\ =1 \ , \ x \in [a,\infty) \end{cases}$$

Equation to Figure 6.3.2.2.3   **Equation to Figure 6.3.2.2.4**

Given the appropriately relevant fuzzy numbers, the conflicts in fuzzy continuum assessments of any epistemic object are that, higher decision value is attached to an item if the item is viewed as positive in the decision-choice process. Similarly, low evaluative decision value is attached if the same item is viewed as negative relative to the phenomenon. Such a principle of behavior in duality helps to overcome the problem of overvaluations and under-valuations in the true-false claims. The concepts of positive and negative are linguistic variables with qualitative characteristics whose values are subjectively defined and assessed by corresponding membership functions that present a spectrum of fuzzy numbers in a continuum under the guidance of the toolbox of the fuzzy paradigm. The strength of the fuzzy paradigm in the knowledge-production process under decision-choice rationalities is to be found in the three elements of representation of inexact symbolic ideas and propositions, the principle of continuum of opposites, the acceptance of contradiction as a valid truth value and the fuzzy laws of thought.

The fuzzy logic is characterized by duality and continuum of opposites with internal conflicts that produce contradictions that demand judicious judgment by decision-choice agents. The specification of the duality with negative and positive characteristic sets equipped with membership characteristic functions and the resolution of the internal conflict by subjective action lead to fuzzy rationality within the continuum. The resolution is such that the negative characteristic set in the duality acts as a constraint on the positive characteristic set and vice versa. In other words, the judgment of positivity (negativity) is constrained by conditions of negativity (positivity) and cognitively formulated as maximization (minimization) of positivity (negativity) subject to negativity (positivity) as a constraint on the decision-choice action and subjective judgment. This approach, in addition to fuzzy aggregation, provides a toolbox of analytical methods for dealing with defective information structure over the epistemological process to obtain exact-value equivalences in the knowledge-production system. The same fuzzy paradigm provides us with powerful analytical toolbox to accommodate the analysis of deceptive information structure in the socio-physical systems. The approach presented here can be used to deal with the exact-inexact demarcation problem in the information-knowledge space as well as deal with the problem of inexact probabilities and stochastic processes. Finally, it may be added that these fuzzy numbers may be approximated by triangular fuzzy, trapezoidal fuzzy numbers through fuzzy decomposition process.

References

R1. Bounded Rationality in Knowledge Systems

[R1.1] Arthur, W.B.: Designing Economic Agents that Act Like Human Agents: A Behavioral Approach to Bounded Rationality. American Economic Review: Papers and Proceedings 81, 353–359 (1991)

[R1.2] Dow, J.: Search Decisions with Limited Memory. Review of Economic Studies 58, 1–14 (1991)

[R1.3] Gigerenzer, G., Selten, R.: Bounded Rationality: The Adaptive Toolbox. MIT Press, Cambridge (2001)

[R1.4] Gigerenzer, G., Goldstein, D.G.: Reasoning the Fast and Frugal Way: Models of Bounded Rationality. Psychological Review 103, 650–669 (1996)

[R1.5] Gigerenzer, G.: Bounded Rationality: Models of Fast and Frugal Inference. Swiss Journal of Economic Statistics 133, 201–218 (1997)

[R1.6] Honkapohja, S.: Adaptive Learning and Bounded Rationality. European Economic Review 37, 587–594

[R1.7] Lipman, B.: Information Processing and Bounded Rationality: A Survey. Canadian Jour. of Economics 28, 42–63 (1995)

[R1.8] Lipman, B.: How to Decide How to Decide How...: Modeling Limited Rationality. Econometrica 59, 1105–1125 (1991)

[R1.9] March, J.G.: Bounded Rationality, Ambiguity and Engineering of Choice. The Bell Journal of Economics 9(2), 587–608 (1978)

[R1.10] Neyman, A.: Bounded Rationality Justifies Cooperation in the Finitely Repeated Prisons' Dilemma Game. Economic Letters 19, 227–229

[R1.11] Radner, R.: Can Bounded Rationality Resolve the Prisoner's Dilemma? In: Mas-Colell, A., Hildenbrand, W. (eds.) Essays in Honor of Gerard Debreu. North-Holland, Amsterdam (1986)

[R1.12] Rieskamp, J., et al.: Extending the Bounds of Rationality: Evidence and Theories of Preferential Choice. Journal of Economic Literature 44, 631–661 (2006)

[R1.13] Rosenthal, R.: A Bounded-Rationality Approach to the Study of Noncooperative Games. International Journal of Game Theory 18, 273–292 (1989)

[R1.14] Rubinstein, A.: Modeling Bounded Rationality. MIT Press, Cambridge (1998)

[R1.15] Rubinstein, A.: New Directions in Economic Theory – Bounded Rationality. Revista Española de Economia 7, 3–15 (1990)

[R1.16] Sargent, T.: Bounded Rationality in Macroeconomics. Clarendon, Oxford (1993)

[R1.17] Simon, H.A.: Theories of Bounded Rationality. In: McGuire, C.B., et al. (eds.) Decision and Organization, pp. 161–176. North Holland, Amsterdam (1972)

[R1.18] Simon, H.A.: Models of Bounded Rationality, vol. 2. MIT Press, Cambridge (1982)

[R1.19] Simon, H.A.: From Substantive to Procedural Rationality. In: Latis, S.J. (ed.) Methods and Apprasal in Economics, pp. 129–148. Cambridge University Press, New York (1976)

[R1.20] Starbuck, W.H.: Levels of Aspiration. Psychological Review 70, 51–60 (1963)

[R1.21] Zemel, E.: Small Talk and Cooperation: A Note on Bounded Rationality. Journal of Economic Theory 49, 1–9

[R1.22] Stigum, B.P., et al.: Foundation of Utility and Risk Theory with Application. D. Reidel Pub., Boston (1983)

R2. Category Theory in Mathematics, Logic and Sciences

[R2.1] Awodey, S.: Structure in Mathematics and Logic: A Categorical Perspective. Philosophia Mathematica 3, 209–237 (1996)

[R2.2] Bell, J.L.: Category Theory and the Foundations of Mathematics. British Journal of Science 32, 349–358 (1981)

[R2.3] Bell, J.L.: Categories, Toposes and Sets. Synthese 51, 393–337 (1982)

[R2.4] Black, M.: The Nature of Mathematics. Adams and Co., Totowa (1965)

[R2.5] Blass, A.: The Interaction Between Category and Set Theory. Mathematical Applications of Category Theory 30, 5–29 (1984)

[R2.6] Brown, B., Woods, J. (eds.): Logical Consequence; Rival Approaches and New Studies in exact Philosophy: Logic, Mathematics and Science, Oxford, Hermes, vol. II (2000)

[R2.7] Butts, R.: Logic, Foundations of Mathematics and Computability. Reidel, Boston (1977)

[R2.8] Domany, J.L., et al.: Models of Neural Networks III. Springer, New York (1996)

[R2.9] Dompere, K.K.: Fuzzy Rationality: Methodological Critique and Unity of Classical, Bounded and Other Rationalities. Springer, New York (2009)

[R2.10] Dompere, K.K.: The Theory of the Knowledge Square: The Fuzzy Analytical Foundations of Knowing, A Working Monograph on Philosophy of Science I. Howard University, Washington, D.C. (2011)

[R2.11] Dompere, K.K.: Fuzzy Rational Foundations of Exact and Inexact Sciences, A Working Monograph on Philosophy of Science II. Howard University, Washington, D.C. (2011)

[R2.12] Feferman, S.: Categorical Foundations and Foundations of Category Theory. In: Butts, R. (ed.) Logic, Foundations of Mathematics and Computability, pp. 149–169. Reidel, Boston (1977)

[R2.13] Gray, J.W. (ed.): Mathematical Applications of Category Theory (American Mathematical Society Meeting 89th Denver Colo. 1983). American Mathematical Society, Providence (1984)

[R.2.14] Johansson, I.: Ontological Investigations: An Inquiry into the Categories of Nature, Man, and Society. Routledge, New York (1989)

[R2.15] Kamps, K.H., Pumplun, D., Tholen, W. (eds.): Category Theory: Proceedings of the International Conference, Gummersbach, July 6-10. Springer, New York (1982)

[R2.16] Kent, B., Peirce, C.S.: Logic and the Classification of the Sciences. McGill-Queen,s University Press, Kingston (1987)

[R2.17] Kosko, B.: Neural Networks and Fuzzy Systems, Englewood Cliffs, NJ (1991)

[R2.18] Landry, E.: Category Theory: the Language of Mathematics. Philosophy of Science 66, (supp.) S14–S27

[R2.19] Landry, E., Marquis, J.P.: Categories in Context: Historical, Foundational and Philosophical. Philosophia Mathematica 13, 1–43 (2005)

[R2.20] Marquis, J.-P.: Three Kinds of Universals in Mathematics. In: Brown, B., Woods, J. (eds.) Logical Consequence; Rival Approaches and New Studies in exact Philosophy: Logic, Mathematics and Science, Oxford, Hermes, vol. II, pp. 191–212 (2000)

[R2.21] McLarty, C.: Category Theory in Real Time. Philosophia Mathematica 2, 36–44 (1994)

[R2.22] McLarty, C.: Learning from Questions on Categorical Foundations. Philosophia Mathematica 13, 44–60 (2005)

[R2.23] McLarty, C.: Learning from Questions on Categorical Foundations. Philosophia Mathematica 13, 44–60 (2005)

[R2.24] Taylor, J.G. (ed.): Mathematical Approaches to Neural Networks. North-Holland, New York (1993)

[R2.25] Van Benthem, J., et al. (eds.): The Age of Alternative Logics: Assessing Philosophy of Logic and Mathematics Today. Springer, New York (2006)

R3. Fuzzy Logic in Knowledge Production

[R3.1]   Baldwin, J.F.: A New Approach to Approximate Reasoning Using a Fuzzy Logic. Fuzzy Sets and Systems 2(4), 309–325 (1979)

[R3.2]   Baldwin, J.F.: Fuzzy Logic and Fuzzy Reasoning. Intern. J. Man-Machine Stud. 11, 465–480 (1979)

[R3.3]   Baldwin, J.F.: Fuzzy Logic and Its Application to Fuzzy Reasoning. In: Gupta, M.M., et al. (eds.) Advances in Fuzzy Set Theory and Applications, pp. 96–115. North-Holland, New York (1979)

[R3.4]   Baldwin, J.F., et al.: Fuzzy Relational Inference Language. Fuzzy Sets and Systems 14(2), 155–174 (1984)

[R3.5]   Baldsin, J., Pilsworth, B.W.: Axiomatic Approach to Implication For Approximate Reasoning With Fuzzy Logic. Fuzzy Sets and Systems 3(2), 193–219 (1980)

[R3.6]   Baldwin, J.F., et al.: The Resolution of Two Paradoxes by Approximate Reasoning Using A Fuzzy Logic. Synthese 44, 397–420 (1980)

[R3.7]   Fukami, S., et al.: Some Considerations On Fuzzy Conditional Inference. Fuzzy Sets and Systems 4(3), 243–273 (1980)

[R3.8]   Gaines, B.R.: Fuzzy Reasoning and the Logic of Uncertainty. In: Proc. 6th International Symp. of Multiple-Valued Logic, IEEE 76CH 1111-4C, pp. 179–188 (1976)

[R3.9]   Gaines, B.R.: Foundations of Fuzzy Reasoning. Inter. Jour. of Man-Machine Studies 8, 623–668 (1976)

[R3.10] Gaines, B.R.: Foundations of Fuzzy Reasoning. In: Gupta, M.M., et al. (eds.) Fuzzy Information and Decision Processes, pp. 19–75. North-Holland, New York (1982)

[R3.11] Gaines, B.R.: Precise Past, Fuzzy Future. International Journal. of Man-Machine Studies 19(1), 117–134 (1983)

[R3.12] Gains, B.R.: Fuzzy and Probability Uncertainty Logics. Information and Control 38, 154–169 (1978)

[R3.13] Gains, B.R.: Modeling Practical Reasoning. Intern. Jour. of Intelligent Systems 8(1), 51–70 (1993)

[R3.14] Gaines, B.R.: Łukasiewicz Logic and Fuzzy Set Theory. International Jour. of Man-Machine Studies 8, 313–327 (1976)

[R3.15] Giles, R.: Lukasiewics Logic and Fuzzy Set Theory. Intern. J. Man-Machine Stud. 8, 313–327 (1976)

[R3.16] Giles, R.: Formal System for Fuzzy Reasoning. Fuzzy Sets and Systems 2(3), 233–257 (1979)

[R3.17] Ginsberg, M.L. (ed.): Readings in Non-monotonic Reason. Morgan Kaufmann, Los Altos (1987)

[R3.18] Goguen, J.A.: The Logic of Inexact Concepts. Synthese 19, 325–373 (1969)

[R3.19] Gottinger, H.W.: Towards a Fuzzy Reasoning in the Behavioral Science. Cybernetica 16(2), 113–135 (1973)

[R3.20] Gottinger, H.W.: Some Basic Issues Connected With Fuzzy Analysis. In: Klaczro, H., Muller, N. (eds.) Systems Theory in Social Sciences, pp. 323–325. Birkhauser Verlag, Basel (1976)

[R3.21] Gottwald, S.: Fuzzy Propositional Logics. Fuzzy Sets and Systems 3(2), 181–192 (1980)

[R3.22] Gupta, M.M., et al. (eds.): Approximate Reasoning In Decision Analysis. North Holland, New York (1982)

[R3.23] Ulrich, H., Klement, E.P.: Non-Clasical Logics and their Applications to Fuzzy Subsets: A Handbook of the Mathematical Foundations of Fuzzy Set Theory. Kluwer, Boston (1995)

[R3.24] Kaipov, V.K., et al.: Classification in Fuzzy Environments. In: Gupta, M.M., et al. (eds.) Advances in Fuzzy Set Theory and Applications, pp. 119–124. North-Holland, New York (1979)

[R3.25] Kaufman, A.: Progress in Modeling of Human Reasoning of Fuzzy Logic. In: Gupta, M.M., et al. (eds.) Fuzzy Information and Decision Process, pp. 11–17. North-Holland, New York (1982)

[R3.26] Lakoff, G.: Hedges: A Study in Meaning Criteria and the Logic of Fuzzy Concepts. Jour. Philos. Logic 2, 458–508 (1973)

[R3.27] Lee, E.T., et al.: Some Properties of Fuzzy Logic. Information and Control 19, 417–431 (1971)

[R3.28] Lee, R.C.T.: Fuzzy Logic and the Resolution Principle. Jour. of Assoc. Comput. Mach. 19, 109–119 (1972)

[R3.29] LeFaivre, R.A.: The Representation of Fuzzy Knowledge. Jour. of Cybernetics 4, 57–66 (1974)

[R3.30] Mitra, S., Pal, S.K.: Logical Operation based Fuzzy MLP for Classification and Rule Generation. Neural Networks 7(2), 353–373 (1994)

[R3.31] Mizumoto, M.: Fuzzy Conditional Inference Under Max- ⊙ Composition. Information Sciences 27(3), 183–207 (1982)

[R3.32] Mizumoto, M., et al.: Several Methods for Fuzzy Conditional Inference. In: Proc. of IEEE Conf. on Decision and Control, Florida, Florida, December 12-14, pp. 777–782 (1979)

[R3.33] Montero, F.J.: Measuring the Rationality of a Fuzzy Preference Relation. Busefal. 26, 75–83 (1986)

[R3.34] Morgan, C.G.: Methods for Automated Theorem Proving in Non-Classical Logics. IEEE Trans. Compt. C-25, 852–862 (1976)

[R3.35] Negoita, C.V.: Representation Theorems for Fuzzy Concepts. Kybernetes 4, 169–174 (1975)

[R3.36] Nguyen, H.T., Walker, E.A.: A First Cours in Fuzzy Logic. CRC Press, Boca Raton (1997)

[R3.37] Nowakowska, M.: Methodological Problems of Measurements of Fuzzy Concepts in Social Sciences. Behavioral Sciences 22(2), 107–115 (1977)

[R3.38] Pinkava, V.: Fuzzification of Binary and Finite Multivalued Logical Calculi. Intern. Jour. Man-Machine Stud. 8, 171–730 (1976)

[R3.39] Skala, H.J.: Non-Archimedean Utility Theory. D. Reidel, Dordrecht (1975)

[R3.40] Skala, H.J.: On Many-Valued Logics, Fuzzy Sets, Fuzzy Logics and Their Applications. Fuzzy Sets and Systems 1(2), 129–149 (1978)

[R3.41] Sugeno, M., Takagi, T.: Multi-Dimensional Fuzzy Reasoning. Fuzzy Sets and Systems 9(3), 313–325 (1983)

[R3.42] Tan, S.K., et al.: Fuzzy Inference Relation Based on the Theory of Falling Shadows. Fuzzy Sets and Systems 53(2), 179–188 (1993)

[R3.43] Thornber, K.K.: A Key to Fuzzy Logic Inference. Intern. Jour. of Approximate Reasoning 8(2), 105–129 (1993)

[R3.44] Tong, R.M., et al.: A Critical Assessment of Truth Functional Modification and its Use in Approximate Reasoning. Fuzzy Sets and Systems 7(1), 103–108 (1982)

[R3.45] Van Fraassen, B.C.: Comments: Lakoff's Fuzzy Propositional Logic. In: Hockney, D., et al. (eds.) Contemporary Research in Philosophical Logic and Linguistic Semantics Holland, pp. 273–277. Reild (1975)

[R3.46] Whalen, T., et al.: Usuality, Regularity, and Fuzzy Set Logic. Intern. Jour. of Approximate Reasoning 6(4), 481–504 (1992)

[R3.47] Yager, R.R., et al. (eds.): An Introduction to Fuzzy Logic Applications in Intelligent Systems. Kluwer, Boston (1992)

[R3.48] Yan, J., et al.: Using Fuzzy Logic: Towards Intelligent Systems. Prentice Hall, Englewood Cliffs (1994)

[R3.49] Ying, M.S.: Some Notes on Multidimensional Fuzzy Reasoning. Cybernetics and Systems 19(4), 281–293 (1988)

[R3.50] Zadeh, L.A.: Quantitative Fuzzy Semantics. Inform. Science 3, 159–176 (1971)

[R3.51] Zadeh, L.A.: A Fuzzy Set Interpretation of Linguistic Hedges. Jour. Cybernetics 2, 4–34 (1972)

[R3.52] Zadeh, L.A.: Fuzzy Logic and Its Application to Approximate Reasoning. In: Information Processing 74. Proc. IFIP Congress, vol. 74(3), pp. 591–594. North Holland, New York (1974)

[R3.53] Zadeh, L.A.: The Concept of a Linguistic Variable and Its Application to Approximate Reasoning. In: Fu, K.S., et al. (eds.) Learning Systems and Intelligent Robots, pp. 1–10. Plenum Press, New York (1974)

[R3.54] Zadeh, L.A.: Fuzzy Logic and Approximate Reasoning. Syntheses 30, 407–428 (1975)

[R3.55] Zadeh, L.A., et al. (eds.): Fuzzy Logic for the Management of Uncertainty. Wily and Sons, New York (1992)

[R3.56] Zadeh, L.A., et al. (eds.): Fuzzy Sets and Their Applications to Cognitive and Decision Processes. Academic Press, New York (1974)

[R3.57] Zadeh, L.A.: The Birth and Evolution of Fuzzy Logic. Intern. Jour. of General Systems 17(2-3), 95–105 (1990)

R4. Fuzzy Mathematics in Approximate Reasoning under Conditions of Inexactness and Vagueness

[R4.1]   Bandler, W., et al.: Fuzzy Power Sets and Fuzzy Implication Operators. Fuzzy Sets and Systems 4(1), 13–30 (1980)

[R4.2]   Banon, G.: Distinction between Several Subsets of Fuzzy Measures. Fuzzy Sets and Systems 5(3), 291–305 (1981)

[R4.3]   Bellman, R.E.: Mathematics and Human Sciences. In: Wilkinson, J., et al. (eds.) The Dynamic Programming of Human Systems, pp. 11–18. MSS Information Corp., New York (1973)

[R4.4]   Bellman, R.E., Glertz, M.: On the Analytic Formalism of the Theory of Fuzzy Sets. Information Science 5, 149–156 (1973)

[R4.5]   Buckley, J.J.: The Fuzzy Mathematics of Finance. Fuzzy Sets and Systems 21(3), 257–273 (1987)

[R4.6]   Butnariu, D.: Fixed Points for Fuzzy Mapping. Fuzzy Sets and Systems 7(2), 191–207 (1982)

[R4.7]   Butnariu, D.: Decompositions and Range For Additive Fuzzy Measures. Fuzzy Sets and Systems 10(2), 135–155 (1983)

[R4.8]   Cerruti, U.: Graphs and Fuzzy Graphs. In: Fuzzy Information and Decision Processes, pp. 123–131. North-Holland, Amsterdam (1982)

[R4.9]   Chakraborty, M.K., et al.: Studies in Fuzzy Relations Over Fuzzy Subsets. Fuzzy Sets and Systems 9(1), 79–89 (1983)

[R4.10] Chang, C.L.: Fuzzy Topological Spaces. J. Math. Anal. and Applications 24, 182–190 (1968)

[R4.11] Chang, S.S.L.: Fuzzy Mathematics, Man and His Environment. IEEE Transactions on Systems, Man and Cybernetics, SMC-S 2, 92–93 (1972)

[R4.12] Chang, S.S.L., et al.: On Fuzzy Mathematics and Control. IEEE Transactions, System, Man and Cybernetics SMC-2, 30–34 (1972)

[R4.13] Chang, S.S.: Fixed Point Theorems for Fuzzy Mappings. Fuzzy Sets and Systems 17, 181–187 (1985)

[R4.14] Chapin, E.W.: An Axiomatization of the Set Theory of Zadeh. Notices. American Math. Society 687-02-4 754 (1971)

[R4.15] Chaudhury, A.K., Das, P.: Some Results on Fuzzy Topology on Fuzzy Sets. Fuzzy Sets and Systems 56, 331–336 (1993)

[R4.16] Cheng-Zhong, L.: Generalized Inverses of Fuzzy Matrix. In: Gupta, M.M., et al. (eds.) Approximate Reasoning In Decision Analysis, pp. 57–60. North-Holland, Amsterdam (1982)

[R4.17] Chitra, H., Subrahmanyam, P.V.: Fuzzy Sets and Fixed Points. Jour. of Mathematical Analysis and Application 124, 584–590 (1987)

[R4.18] Cohn, D.L.: Measure Theory. Birkhauser, Boston (1980)

[R4.19] Cohen, P.J., Hirsch, R.: Non-Cantorian Set Theory. Scientific America, 101–116 (December 1967)

[R4.20] Czogala, J., et al.: Fuzzy Relation Equations On a Finite Set. Fuzzy Sets and Systems 7(1), 89–101 (1982)

[R4.21] Das, P.: Fuzzy Topology on Fuzzy Sets: Product Fuzzy Topology and Fuzzy Topological Groups. Fuzzy Sets and Systems 100, 367–372 (1998)

[R4.22] DiNola, A., et al. (eds.): The Mathematics of Fuzzy Systems. Verlag TUV Rheinland, Koln (1986)

[R4.23] Dombi, J.: A General Class of Fuzzy Operators, the DeMorgan Class of Fuzzy Operators and Fuzzy Measures Induced by Fuzzy Operators. Fuzzy Sets and Systems 8(2), 149–163 (1982)

[R4.24] Dubois, D., Prade, H.: Fuzzy Sets and Systems. Academic Press, New York (1980)

[R4.25] Dubois: Fuzzy Real Algebra: Some Results. Fuzzy Sets and Systems 2(4), 327–348 (1979)

[R4.26] Dubois, D., Prade, H.: Gradual Inference Rules in Approximate Reasoning. Information Sciences 61(1-2), 103–122 (1992)

[R4.27] Dubois, D., Prade, H.: On the Combination of Evidence in various Mathematical Frameworks. In: Flamm, J., Luisi, T. (eds.) Reliability Data Collection and Analysis, pp. 213–241. Kluwer, Boston (1992)

[R4.28] Dubois, D., Prade, H.: Fuzzy Sets and Probability: Misunderstanding, Bridges and Gaps. In: Proc. Second IEEE Intern. Conf. on Fuzzy Systems, San Francisco, pp. 1059–1068 (1993)

[R4.29] Dubois, D., Prade, H.: A Survey of Belief Revision and Updating Rules in Various Uncertainty Models. Intern. J. of Intelligent Systems 9(1), 61–100 (1994)

[R4.30] Erceg, M.A.: Functions, Equivalence Relations, Quotient Spaces and Subsets in Fuzzy Set Theory. Fuzzy Sets and Systems 3(1), 79–92 (1980)

[R4.31] Feng, Y.-J.: A Method Using Fuzzy Mathematics to Solve the Vectormaximum Problem. Fuzzy Sets and Systems 9(2), 129–136 (1983)

[R4.32] Filev, D.P., et al.: A Generalized Defuzzification Method via Bag Distributions. Intern. Jour. of Intelligent Systems 6(7), 687–697 (1991)

[R4.33] Foster, D.H.: Fuzzy Topological Groups. Journal of Math. Analysis and Applications 67, 549–564 (1979)

[R4.34] Goetschel Jr., R., et al.: Topological Properties of Fuzzy Number. Fuzzy Sets and Systems 10(1), 87–99 (1983)

[R4.35] Goodman, I.R.: Fuzzy Sets As Random Level Sets: Implications and Extensions of the Basic Results. In: Lasker, G.E. (ed.) Applied Systems and Cybernetics: Fuzzy Sets and Systems, vol. VI, pp. 2756–2766. Pergamon Press, New York (1981)

[R4.36] Goodman, I.R.: Fuzzy Sets As Equivalence Classes of Random Sets. In: Yager, R.R. (ed.) Fuzzy Set and Possibility Theory: Recent Development, pp. 327–343. Pergamon Press, Oxford (1992)

[R4.37] Gupta, M.M., et al. (eds.): Fuzzy Antomata and Decision Processes. North-Holland, New York (1977)

[R4.38] Gupta, M.M., Sanchez, E. (eds.): Fuzzy Information and Decision Processes. North-Holland, New York (1982)

[R4.39] Higashi, M., Klir, G.J.: On measure of fuzziness and fuzzy complements. Intern. J. of General Systems 8(3), 169–180 (1982)

[R4.40] Higashi, M., Klir, G.J.: Measures of uncertainty and information based on possibility distributions. International Journal of General Systems 9(1), 43–58 (1983)

[R4.41] Higashi, M., Klir, G.J.: On the notion of distance representing information closeness: Possibility and probability distributions. Intern. J. of General Systems 9(2), 103–115 (1983)

[R4.42] Higashi, M., Klir, G.J.: Resolution of finite fuzzy relation equations. Fuzzy Sets and Systems 13(1), 65–82 (1984)

[R4.43] Higashi, M., Klir, G.J.: Identification of fuzzy relation systems. IEEE Trans. on Systems, Man, and Cybernetics 14(2), 349–355 (1984)

[R4.44] Ulrich, H.: A Mathematical Theory of Uncertainty. In: Yager, R.R. (ed.) Fuzzy Set and Possibility Theory: Recent Developments, pp. 344–355. Pergamon, New York (1982)

[R4.45] Jin-wen, Z.: A Unified Treatment of Fuzzy Set Theory and Boolean Valued Set theory: Fuzzy Set Structures and Normal Fuzzy Set Structures. Jour. Math. Anal. and Applications 76(1), 197–301 (1980)

[R4.46] Kandel, A.: Fuzzy Mathematical Techniques with Applications. Addison-Wesley, Reading (1986)

[R4.47] Kandel, A., Byatt, W.J.: Fuzzy Processes. Fuzzy Sets and Systems 4(2), 117–152 (1980)

[R4.48] Kaufmann, A., Gupta, M.M.: Introduction to fuzzy arithmetic: Theory and applications. Van Nostrand Rheinhold, New York (1991)

[R4.49] Kaufmann, A.: Introduction to the Theory of Fuzzy Subsets, vol. 1. Academic Press, London (1975)

[R4.50] Kaufmann, A.: Theory of Fuzzy Sets. Merson Press, Paris (1972)

[R4.51] Kaufmann, A., et al.: Fuzzy Mathematical Models in Engineering and Management Science. North-Holland, New York (1988)

[R4.52] Kim, K.H., et al.: Generalized Fuzzy Matrices. Fuzzy Sets and Systems 4(3), 293–315 (1980)

[R4.53] Klement, E.P.: Fuzzy and σ – Algebras Fuzzy Measurable Functions. Fuzzy Sets and Systems 4, 83–93 (1980)

[R4.54] Klement, E.P.: Characterization of Finite Fuzzy Measures Using Markoff-kernels. Journal of Math. Analysis and Applications 75, 330–339 (1980)

[R4.55] Klement, E.P.: Construction of Fuzzy σ – Algebras Using Triangular Norms. Journal of Math. Analysis and Applications 85, 543–565 (1982)

[R4.56] Klement, E.P., Schwyhla, W.: Correspondence Between Fuzzy Measures and Classical Measures. Fuzzy Sets and Systems 7(1), 57–70 (1982)

[R4.57] Klir, G., Yuan, B.: Fuzzy Sets and Fuzzy Logic. Prentice Hall, Upper Saddle River (1995)

[R4.58] Kokawa, M., et al.: Fuzzy-Theoretical Dimensionality Reduction Method of Multi-Dimensional Quality. In: Gupta, M.M., Sanchez, E. (eds.) Fuzzy Information and Decision Processes, pp. 235–250. North-Holland, New York (1982)

[R4.59] Kramosil, I., et al.: Fuzzy Metrics and Statistical Metric Spaces. Kybernetika 11, 336–344 (1975)

[R4.60] Kruse, R.: On the Construction of Fuzzy Measures. Fuzzy Sets and Systems 8(3), 323–327 (1982)

[R4.61] Kruse, R., et al.: Foundations of Fuzzy Systems. John Wiley and Sons, New York (1994)

[R4.62] Lasker, G.E. (ed.): Applied Systems and Cybernetics. Fuzzy Sets and Systems, vol. VI. Pergamon Press, New York (1981)

[R4.63] Lientz, B.P.: On Time Dependent Fuzzy Sets. Inform., Science 4, 367–376 (1972)

[R4.64] Lowen, R.: Fuzzy Uniform Spaces. Jour. Math. Anal. Appl. 82(21981), 367–376 (1981)

[R4.65] Lowen, R.: On the Existence of Natural Non-Topological Fuzzy Topological Space. Haldermann Verlag, Berlin (1986)

[R4.66] Martin, H.W.: Weakly Induced Fuzzy Topological Spaces. Jour. Math. Anal.
         and Application 78, 634–639 (1980)
[R4.67] Michalek, J.: Fuzzy Topologies. Kybernetika 11, 345–354 (1975)
[R4.68] Mizumoto, M., Tanaka, K.: Some Properties of Fuzzy Numbers. In: Gupta,
         M.M., et al. (eds.) Advances in Fuzzy Sets Theory and Applications, pp. 153–
         164. North-Holland, Amsterdam (1979)
[R4.69] Negoita, C.V., et al.: Applications of Fuzzy Sets to Systems Analysis. Wiley and
         Sons, New York (1975)
[R4.70] Negoita, C.V.: Representation Theorems for Fuzzy Concepts. Kybernetes 4,
         169–174 (1975)
[R4.71] Negoita, C.V., et al.: On the State Equation of Fuzzy Systems. Kybernetes 4,
         213–214 (1975)
[R4.72] Negoita, C.V.: Fuzzy Sets in Topoi. Fuzzy Sets and Systems 8(1), 93–99 (1982)
[R4.73] Netto, A.B.: Fuzzy Classes. Notices, American Mathematical Society 68T-H28,
         945 (1968)
[R4.74] Nguyen, H.T.: Possibility Measures and Related Topics. In: Gupta, M.M., et al.
         (eds.) Approximate Reasoning In Decision Analysis, pp. 197–202. North
         Holland, New York (1982)
[R4.75] Nowakowska, M.: Some Problems in the Foundations of Fuzzy Set Theory. In:
         Gupta, M.M., et al. (eds.) Approximate Reasoning In Decision Analysis, pp.
         349–360. North Holland, New York (1982)
[R4.76] Ovchinnikov, S.V.: Structure of Fuzzy Binary Relations. Fuzzy Sets and
         Systems 6(2), 169–195 (1981)
[R4.77] Pedrycz, W.: Fuzzy Relational Equations with Generalized Connectives and
         Their Applications. Fuzzy Sets and Systems 10(2), 185–201 (1983)
[R4.78] Raha, S., et al.: Analogy Between Approximate Reasoning and the Method of
         Interpolation. Fuzzy Sets and Systems 51(3), 259–266 (1992)
[R4.79] Ralescu, D.: Toward a General Theory of Fuzzy Variables. Jour. of Math.
         Analysis and Applications 86(1), 176–193 (1982)
[R4.80] Rao, M.B., et al.: Some Comments On Fuzzy Variables. Fuzzy Sets and
         Systems 6(2), 285–292 (1981)
[R4.81] Rodabaugh, S.E.: Fuzzy Arithmetic and Fuzzy Topology. In: Lasker, G.E. (ed.)
         Applied Systems and Cybernetics. Fuzzy Sets and Systems, vol. VI, pp. 2803–
         2807. Pergamon Press, New York (1981)
[R4.82] Rodabaugh, S., et al. (eds.): Application of Category Theory to Fuzzy Subsets.
         Kluwer, Boston (1992)
[R4.83] Roubens, M., et al.: Linear Fuzzy Graphs. Fuzzy Sets and Systems 10(1), 79–86
         (1983)
[R4.84] Rosenfeld, A.: Fuzzy Groups. Jour. Math. Anal. Appln. 35, 512–517 (1971)
[R4.85] Rosenfeld, A.: Fuzzy Graphs. In: Zadeh, L.A., et al. (eds.) Fuzzy Sets and Their
         Applications to Cognitive and Decision Processes, pp. 77–95. Academic Press,
         New York (1974)
[R4.86] Ruspini, E.H.: Recent Developments In Mathematical Classification Using
         Fuzzy Sets. In: Lasker, G.E. (ed.) Applied Systems and Cybernetics. Fuzzy Sets
         and Systems, vol. VI, pp. 2785–2790. Pergamon Press, New York (1981)
[R4.87] Santos, E.S.: Maximin, Minimax and Composite Sequential Machines. Jour.
         Math. Anal. and Appln. 24, 246–259 (1968)
[R4.88] Santos, E.S.: Fuzzy Algorithms. Inform. and Control 17, 326–339 (1970)

[R4.89]   Sarkar, M.: On Fuzzy Topological Spaces. Jour. Math. Anal. Appln. 79, 384–394 (1981)

[R4.90]   Slowinski, R., Teghem, J. (eds.): Stochastic versus Fuzzy Approaches to Multiobjective Mathematical Programming Under Uncertainty. Kluwer, Dordrecht (1990)

[R4.91]   Stein, N.E., Talaki, K.: Convex Fuzzy Random Variables. Fuzzy Sets and Systems 6(3), 271–284 (1981)

[R4.92]   Sugeno, M.: Inverse Operation of Fuzzy Integrals and Conditional Fuzzy Measures. Transactions SICE 11, 709–714 (1975)

[R4.93]   Yager, R.R., Filver, D.P.: Essentials of Fuzzy Modeling and Control. John Wiley and Sons, New York (1994)

[R4.94]   Triantaphyllon, E., et al.: The Problem of Determining Membership Values in Fuzzy Sets in Real World Situations. In: Brown, D.E., et al. (eds.) Operations Research and Artificial Intelligence: The Integration of Problem-Solving Strategies, pp. 197–214. Kluwer, Boston (1990)

[R4.95]   Tsichritzis, D.: Participation Measures. Jour. Math. Anal. and Appln. 36, 60–72 (1971)

[R4.96]   Tsichritzis, D.: Approximation and Complexity of Functions on the Integers. Inform. Science 4, 70–86 (1971)

[R4.97]   Turksens, I.B.: Four Methods of Approximate Reasoning with Interval-Valued Fuzzy Sets. Intern. Journ. of Approximate Reasoning 3(2), 121–142 (1989)

[R4.98]   Turksen, I.B.: Measurement of Membership Functions and Their Acquisition. Fuzzy Sets and Systems 40(1), 5–38 (1991)

[R4.99]   Wang, L.X.: Adaptive Fuzzy Sets and Control: Design and Stability Analysis. Prentice Hall, Englewood Cliffs (1994)

[R4.100] Wang, P.P. (ed.): Advances in Fuzzy Sets, Possibility Theory, and Applications. Plenum Press, New York (1983)

[R4.101] Wang, P.P. (ed.): Advances in Fuzzy Theory and Technology, vol. 1. Bookwright Press, Durham (1992)

[R4.102] Wang, Z., Klir, G.: Fuzzy Measure Theory. Plenum Press, New York (1992)

[R4.103] Wang, P.Z., et al. (eds.): Between Mind and Computer: Fuzzy Science and Engineering. World Scientific Press, Singapore (1993)

[R4.104] Wang, P.Z.: Contactability and Fuzzy Variables. Fuzzy Sets and Systems 8(1), 81–92 (1982)

[R4.105] Wang, S.: Generating Fuzzy Membership Functions: A Monotonic Neural Network Model. Fuzzy Sets and Systems 61(1), 71–82 (1994)

[R4.106] Wierzchon, S.T.: An Algorithm for Identification of Fuzzy Measure. Fuzzy Sets and Systems 9(1), 69–78 (1983)

[R4.107] Wong, C.K.: Fuzzy Topology: Product and Quotient Theorems. Journal of Math. Analysis and Applications 45, 512–521 (1974)

[R4.108] Wong, C.K.: Fuzzy Points and Local Properties of Fuzzy Topology. Jour. Math. Anal. and Appln. 46, 316–328 (1987)

[R4.109] Wong, C.K.: Categories of Fuzzy Sets and Fuzzy Topological Spaces. Jour. Math. Anal. and Appln. 53, 704–714 (1976)

[R4.110] Yager, R.R.: On the Lack of Inverses in Fuzzy Arithmetic. Fuzzy Sets and Systems 4(1), 73–82 (1980)

[R4.111] Yager, R.R. (ed.): Fuzzy Set and Possibility Theory: Recent Development. Pergamon Press, New York (1992)

[R4.112] Yager, R.R.: Fuzzy Subsets with Uncertain Membership Grades. IEEE Transactions on Systems, Man and Cybernetics 14(2), 271–275 (1984)

[R4.113] Yager, R.R., et al. (eds.): Fuzzy Sets, Neural Networks, and Soft Computing. Nostrand Reinhold, New York (1994)

[R4.114] Yager, R.R.: On the Theory of Fuzzy Bags. Intern. Jour. of General Systems 13(1), 23–37 (1986)

[R4.115] Yager, R.R.: Cardinality of Fuzzy Sets via Bags. Mathematical Modelling 9(6), 441–446 (1987)

[R4.116] Zadeh, L.A.: A Computational Theory of Decompositions. Intern. Jour. of Intelligent Systems 2(1), 39–63 (1987)

[R4.117] Zadeh, L.A., et al.: Fuzzy Logic for the Management of Uncertainty. John Wiley, New York (1992)

[R4.118] Zimmerman, H.J.: Fuzzy Set Theory and Its Applications. Kluwer, Boston (1985)

R5. Fuzzy Optimization, Decision-Choices and Approximate Reasoning in Sciences

[R5.1] Bose, R.K., Sahani, D.: Fuzzy Mappings and Fixed Point Theorems. Fuzzy Sets and Systems 21, 53–58 (1987)

[R5.2] Buckley, J.J.: Fuzzy Programming And the Pareto Optimal Set. Fuzzy Set and Systems 10(1), 57–63 (1983)

[R5.3] Butnariu, D.: Fixed Points for Fuzzy Mappings. Fuzzy Sets and Systems 7, 191–207 (1982)

[R5.4] Carlsson, G.: Solving Ill-Structured Problems Through Well Structured Fuzzy Programming. In: Brans, J.P. (ed.) Operation Research 1981, pp. 467–477. North-Holland, Amsterdam (1981)

[R5.5] Carlsson, C.: Tackling an AMDM – Problems with the Help of Some Results From Fuzzy Set Theory. European Journal of Operational Research 10(3), 270–281 (1982)

[R5.6] Cerny, M.: Fuzzy Approach to Vector Optimization. Intern. Jour. of General Systems 20(1), 23–29

[R5.7] Chang, S.S.: Fixed Point Theorems for Fuzzy Mappings. Fuzzy Sets and Systems 17, 181–187 (1985)

[R5.8] Chang, S.Y., et al.: Modeling To Generate Alternatives: A Fuzzy Approach. Fuzzy Sets and Systems 9(2), 137–151 (1983)

[R5.9] Dubois, D.: An Application of Fuzzy Arithmetic to the Optimization of Industrial Machining Processes. Mathematical Modelling 9(6), 461–475 (1987)

[R5.10] Edwards, W.: The Theory of Decision Making. Psychological Bulletin 51, 380–417 (1954)

[R5.11] Eaves, B.C.: Computing Kakutani Fixed Points. Journal of Applied Mathematics 21, 236–244 (1971)

[R5.12] Feng, Y.J.: A Method Using Fuzzy Mathematics to Solve the Vector Maxim Problem. Fuzzy Set and Systems 9(2), 129–136 (1983)

[R5.13] Hannan, E.L.: Linear Programming with Multiple Fuzzy Goals. Fuzzy Sets and Systems 6(3), 235–248 (1981)

[R5.14] Heilpern, S.: Fuzzy Mappings and Fixed Point Theorem. Journal of Mathematical Analysis and Applications 83, 566–569 (1981)

[R.15]   Ignizio, J.P., et al.: Fuzzy Multicriteria Integer Programming via Fuzzy Generalized Networks. Fuzzy Sets and Systems 10(3), 261–270 (1983)

[R5.16] Kacprzyk, J., et al. (eds.): Optimization Models Using Fuzzy Sets and Possibility Theory. D. Reidel, Boston (1987)

[R5.17] Kakutani, S.: A Generalization of Brouwer's Fixed Point Theorem. Duke Mathematical Journal 8, 416–427 (1941)

[R5.18] Kaleva, O.: A Note on Fixed Points for Fuzzy Mappings. Fuzzy Sets and Systems 15, 99–100 (1985)

[R5.19] Kandel, A.: On Minimization of Fuzzy Functions. IEEE Trans. Comp. C 22, 826–832 (1973)

[R5.20] Kandel, A.: On the Minimization of Incompletely Specified Fuzzy Functions. Information, and Control 26, 141–153 (1974)

[R5.21] Lai, Y., et al.: Fuzzy Mathematical Programming. Springer, New York (1992)

[R9.22] Leberling, H.: On Finding Compromise Solution in Multcriteria Problems, Using the Fuzzy Min-Operator. Fuzzy Set and Systems 6(2), 105–118 (1981)

[R5.23] Lee, E.S., et al.: Fuzzy Multiple Objective Programming and Compromise Programming with Pareto Optimum. Fuzzy Sets and Systems 53(3), 275–288 (1993)

[R5.24] Lodwick, W.A., Kacprzyk, J. (eds.): Fuzzy Optimization. STUDFUZZ, vol. 254. Springer, Heidelberg (2010)

[R5.25] Lowen, R.: Convex Fuzzy Sets. Fuzzy Sets and Systems 3, 291–310 (1980)

[R5.26] Luhandjula, M.K.: Compensatory Operators in Fuzzy Linear Programming with Multiple Objectives. Fuzzy Sets and Systems 8(3), 245–252 (1982)

[R5.27] Luhandjula, M.K.: Linear Programming Under Randomness and Fuzziness. Fuzzy Sets and Systems 10(1), 45–54 (1983)

[R5.28] Negoita, C.V., et al.: Fuzzy Linear Programming and Tolerances in Planning. Econ. Group Cybernetic Studies 1, 3–15 (1976)

[R5.29] Negoita, C.V., Stefanescu, A.C.: On Fuzzy Optimization. In: Gupta, M.M., et al. (eds.) Approximate Reasoning In Decision Analysis, pp. 247–250. North Holland, New York (1982)

[R5.30] Negoita, C.V.: The Current Interest in Fuzzy Optimization. Fuzzy Sets and Systems 6(3), 261–270 (1981)

[R5.31] Negoita, C.V., et al.: On Fuzzy Environment in Optimization Problems. In: Rose, J., et al. (eds.) Modern Trends in Cybernetics and Systems, pp. 13–24. Springer, Berlin (1977)

[R5.32] Orlovsky, S.A.: On Formulation of General Fuzzy Mathematical Problem. Fuzzy Sets and Systems 3, 311–321 (1980)

[R5.33] Ostasiewicz, W.: A New Approach to Fuzzy Programming. Fuzzy Sets and Systems 7(2), 139–152 (1982)

[R5.34] Pollatschek, M.A.: Hieranchical Systems and Fuzzy-Set Theory. Kybernetes 6, 147–151 (1977)

[R5.35] Ponsard, G.: Partial Spatial Equilibra With Fuzzy Constraints. Journal of Regional Science 22(2), 159–175 (1982)

[R5.36] Prade, M.: Operations Research with Fuzzy Data. In: Want, P.P., et al. (eds.) Fuzzy Sets, pp. 155–170. Plenum, New York (1980)

[R5.37] Ralescu, D.: Optimization in a Fuzzy Environment. In: Gupta, M.M., et al. (eds.) Advances in Fuzzy Set Theory and Applications, pp. 77–91. North-Holland, New York (1979)

[R5.38] Ralescu, D.A.: Orderings, Preferences and Fuzzy Optimization. In: Rose, J. (ed.) Current Topics in Cybernetics and Systems, Springer, Berlin (1978)

[R5.39] Tanaka, K., et al.: Fuzzy Programs and Their Execution. In: Zadeh, L.A., et al. (eds.) Fuzzy Sets and Their Applications to Cognitive and Decision Processes, pp. 41–76 (1974)

[R5.40] Tanaka, H., et al.: On Fuzzy-Mathematical Programming. Journal of Cybernetics 3(4), 37–46 (1974)

[R5.41] Vira, J.: Fuzzy Expectation Values in Multistage Optimization Problems. Fuzzy Sets and Systems 6(2), 161–168 (1981)

[R5.42] Verdegay, J.L.: Fuzzy Mathematical Programming. In: Gupta, M.M., et al. (eds.) Fuzzy Information and Decision Processes, pp. 231–238. North-Holland, New York (1982)

[R5.43] Warren, R.H.: Optimality in Fuzzy Topological Polysystems. Jour. Math. Anal. 54, 309–315 (1976)

[R5.44] Weiss, M.D.: Fixed Points, Separation and Induced Topologies for Fuzzy Sets. Jour. Math. Anal. and Appln. 50, 142–150 (1975)

[R5.45] Wilkinson, J.: Archetypes, Language, Dynamic Programming and Fuzzy Sets. In: Wilkinson, J., et al. (eds.) The Dynamic Programming of Human Systems, pp. 44–53. Information Corp., MSS New York (1973)

[R5.46] Zadeh, L.A.: The Role of Fuzzy Logic in the Management of Ucertainty in expert Systems. Fuzzy Sets and Systems 11, 199–227 (1983)

[R5.47] Zimmerman, H.-J.: Description and Optimization of Fuzzy Systems. Intern. Jour. Gen. Syst. 2(4), 209–215 (1975)

[R5.48] Zimmerman, H.J.: Applications of Fuzzy Set Theory to Mathematical Programming. Information Science 36(1), 29–58 (1985)

R6. Fuzzy Probability, Fuzzy Random Variable and Random Fuzzy Variable

[R6.1] Bandemer, H.: From Fuzzy Data to Functional Relations. Mathematical Modelling 6, 419–426 (1987)

[R6.2] Bandemer, H., et al.: Fuzzy Data Analysis. Kluwer, Boston (1992)

[R6.3] Kruse, R., et al.: Statistics with Vague Data. D. Reidel Pub. Co., Dordrecht (1987)

[R6.4] Chang, R.L.P., et al.: Applications of Fuzzy Sets in Curve Fitting. Fuzzy Sets and Systems 2(1), 67–74

[R6.5] Chen, S.Q.: Analysis for Multiple Fuzzy Regression. Fuzzy Sets and Systems 25(1), 56–65

[R6.6] Celmins, A.: Multidimensional Least-Squares Fitting of Fuzzy Model. Mathematical Modelling 9(9), 669–690

[R6.7] El Rayes, A.B., et al.: Generalized Possibility Measures. Information Sciences 79, 201–222 (1994)

[R6.8] Dumitrescu, D.: Entropy of a Fuzzy Process. Fuzzy Sets and Systems 55(2), 169–177 (1993)

[R6.9] Delgado, M., et al.: On the Concept of Possibility-Probability Consistency. Fuzzy Sets and Systems 21(3), 311–318 (1987)

[R6.10] Devi, B.B., et al.: Estimation of Fuzzy Memberships from Histograms. Information Sciences 35(1), 43–59 (1985)

[R6.11] Diamond, P.: Fuzzy Least Squares. Information Sciences 46(3), 141–157 (1988)

[R6.12] Dubois, D., et al.: Fuzzy Sets, Probability and Measurement. European Jour. of Operational Research 40(2), 135–154 (1989)

[R6.13] Fruhwirth-Schnatter, S.: On Statistical Inference for Fuzzy Data with Applications to Descriptive Statistics. Fuzzy Sets and Systems 50(2), 143–165 (1992)

[R6.14] Fruhwirth-Schnatter, S.: On Fuzzy Bayesian Inference. Fuzzy Sets and Systems 60(1), 41–58 (1993)

[R6.15] Gaines, B.R.: Fuzzy and Probability Uncertainty Logics. Information and Control 38(2), 154–169 (1978)

[R6.16] Geer, J.F., et al.: Discord in Possibility Theory. International Jour. of General Systems 19, 119–132 (1991)

[R6.17] Geer, J.F., et al.: A Mathematical Analysis of Information-Processing Transformation Between Probabilistic and Possibilistic Formulation of Uncertainty. International Jour. of General Systems 20(2), 14–176 (1992)

[R6.18] Goodman, I.R., et al.: Uncertainty Models for Knowledge Based Systems. North-Holland, New York (1985)

[R6.19] Grabish, M., et al.: Fundamentals of Uncertainty Calculi with Application to Fuzzy Systems. Kluwer, Boston (1994)

[R6.20] Guan, J.W., et al.: Evidence Theory and Its Applications, vol. 1. North-Holland, New York (1991)

[R6.21] Guan, J.W., et al.: Evidence Theory and Its Applications, vol. 2. North-Holland, New York (1992)

[R6.22] Hisdal, E.: Are Grades of Membership Probabilities? Fuzzy Sets and Systems 25(3), 349–356 (1988)

[R6.23] Ulrich, H.: A Mathematical Theory of Uncertainty. In: Yager, R.R. (ed.) Fuzzy Set and Possibility Theory: Recent Developments, pp. 344–355. Pergamon, New York (1982)

[R6.24] Kacprzyk, J., Fedrizzi, M. (eds.): Combining Fuzzy Imprecision with Probabilistic Uncertainty in Decision Making. Plenum Press, New York (1992)

[R6.25] Kacprzyk, J., et al.: Combining Fuzzy Imprecision with Probabilistic Uncertainty in Decision Making. Springer, New York (1988)

[R6.26] Klir, G.J.: Where Do we Stand on Measures of Uncertainty, Ambignity, Fuzziness and the like? Fuzzy Sets and Systems 24(2), 141–160 (1987)

[R6.27] Klir, G.J., et al.: Fuzzy Sets, Uncertainty and Information. Prentice Hall, Englewood Cliff (1988)

[R6.28] Klir, G.J., et al.: Probability-Possibility Transformations: A Comparison. Intern. Jour. of General Systems 21(3), 291–310 (1992)

[R6.29] Kosko, B.: Fuzziness vs Probability. Intern. Jour. of General Systems 17(1-3), 211–240 (1990)

[R6.30] Manton, K.G., et al.: Statistical Applications Using Fuzzy Sets. John Wiley, New York (1994)

[R6.31] Meier, W., et al.: Fuzzy Data Analysis: Methods and Indistrial Applications. Fuzzy Sets and Systems 61(1), 19–28 (1994)

[R6.32] Nakamura, A., et al.: A logic for Fuzzy Data Analysis. Fuzzy Sets and Systems 39(2), 127–132 (1991)

[R6.33] Negoita, C.V., et al.: Simulation, Knowledge-Based Compting and Fuzzy Statistics. Van Nostrand Reinhold, New York (1987)

[R6.34] Nguyen, H.T.: Random Sets and Belief Functions. Jour. of Math. Analysis and Applications 65(3), 531–542 (1978)

[R6.35] Prade, H., et al.: Representation and Combination of Uncertainty with belief Functions and Possibility Measures. Comput. Intell. 4, 244–264 (1988)

[R6.36] Puri, M.L., et al.: Fuzzy Random Variables. Jour. of Mathematical Analysis and Applications 114(2), 409–422 (1986)

[R6.37] Rao, N.B., Rashed, A.: Some Comments on Fuzzy Random Variables. Fuzzy Sets and Systems 6(3), 285–292 (1981)

[R6.38] Sakawa, M., et al.: Multiobjective Fuzzy linear Regression Analysis for Fuzzy Input-Output Data. Fuzzy Sets and Systems 47(2), 173–182 (1992)

[R6.39] Schneider, M., et al.: Properties of the Fuzzy Expected Values and the Fuzzy Expected Interval. Fuzzy Sets and Systems 26(3), 373–385 (1988)

[R6.40] Slowinski, R., Teghem, J. (eds.): Stochastic versus Fuzzy Approaches to Multiobjective Mathematical Programming under Uncertainty. Kluwer, Dordrecht (1990)

[R6.41] Stein, N.E., Talaki, K.: Convex Fuzzy Random Variables. Fuzzy Sets and Systems 6(3), 271–284 (1981)

[R6.42] Sudkamp, T.: On Probability-Possibility Transformations. Fuzzy Sets and Systems 51(1), 73–82 (1992)

[R6.43] Tanaka, H., et al.: Possibilistic Linear Regression Analysis for Fuzzy Data. European Jour. of Operational Research 40(3), 389–396 (1989)

[R6.44] Walley, P.: Statistical Reasoning with Imprecise Probabilities. Chapman and Hall, London (1991)

[R6.45] Wang, G.Y., et al.: The Theory of Fuzzy Stochastic Processes. Fuzzy Sets and Systems 51(2), 161–178 (1992)

[R6.46] Wang, X., et al.: Fuzzy Linear Regression Analysis of Fuzzy Valued Variable. Fuzzy Sets and Systems 36(1), 19

[R6.47] Zadeh, L.A.: Probability Measure of Fuzzy Event. Jour. of Math Analysis and Applications 23, 421–427 (1968)

R7. Ideology and the Knowledge Construction Process

[R7.1] Abercrombie, N., et al.: The Dominant Ideology Thesis. Allen and Unwin, London (1980)

[R7.2] Abercrombie, N.: Class, Structure, and Knowledge: Problems in the Sociology of Knowledge. New York University Press, New York (1980)

[R7.3] Aron, R.: The Opium of the Intellectuals. University Press of America, Lanham (1985)

[R7.4] Aronowitz, S.: Science as Power: Discourse and Ideology in Modern Society. University of Minnesota Press, Minneapolis (1988)

[R7.5] Barinaga, M., Marshall, E.: Confusion on the Cutting Edge. Science 257, 616–625 (1992)

[R7.6] Barnett, R.: Beyond All Reason: Living with Ideology in the University. Society for Research into Higher Education and Open University Press, Philadelphia (2003)

[R7.7] Barth, H.: Truth and Ideology. University of California Press, Berkeley (1976)

[R7.8] Basin, A., Verdie, T.: The Economics of Cultural Transmission and the Dynamics of Preferences. Journal of Economic Theory 97, 298–319 (2001)

[R7.9] Beardsley, P.L.: Redefining Rigor: Ideology and Statistics in Political Inquiry. Sage Publications, Bevery Hills (1980)

[R7.10] Bikhchandani, S., et al.: A Theory of Fads, Fashion, Custom, and Cultural Change. Journal of Political Economy 100, 992–1026 (1992)

[R7.11] Robert, B., Richerson, P.J.: Culture and Evolutionary Process. University of Chicago Press, Chicago (1985)

[R7.12] Buczkowski, P., Klawiter, A.: Theories of Ideology and Ideology of Theories. Rodopi, Amsterdam (1986)

[R7.13] Chomsky, N.: Manufacturing Consent. Pantheo Press, New York (1988)

[R7.14] Chomsky, N.: Problem of Knowledge and Freedom. Collins, Glasgow (1972)

[R7.15] Cole, J.R.: Patterns of Intellectual influence in Scientific Research. Sociology of Education 43, 377–403 (1968)

[R7.16] Cole, J.R.: Social Stratification in Science. University of Chicago Press, Chicago (1973)

[R7.17] Debackere, K., Rappa, M.A.: Institutioal Varations in Problem Choice and Persistence among Scientists in an Emerging Fields. Research Policy 23, 425–441 (1994)

[R7.18] Fraser, C., Gaskell, G. (eds.): The Social Psychological Study of Widespread Beliefs. Clarendon Press, Oxford (1990)

[R7.19] Gieryn, T.F.: Problem Retention and Problem Change in Science. Sociological Inquiry 48, 96–115 (1978)

[R7.20] Harrington Jr., J.E.: The Rigidity of social Systems. Journal of Political Economy 107, 40–64

[R7.21] Hinich, M., Munger, M.: Ideology and the Theory of Political Choice. University of Michigan Press, Ann Arbor (1994)

[R7.22] Hull, D.L.: Science as a Process: An Evolutionary Account of the Social and Conceptual Development of Science. University of Chicago Press, Chicago (1988)

[R7.23] Mészáros, I.: Philosophy, Ideology and Social Science: Essay in Negation and Affirmation. Wheatsheaf, Brighton (1986)

[R7.24] Mészáros, I.: The Power of Ideology. New York University Press, New York (1989)

[R7.25] Newcomb, T.M., et al.: Persistence and Change. John Wiley, New York (1967)

[R7.26] Pickering, A.: Science as Practice and Culture. University of Chicago Press, Chicago (1992)

[R7.27] Therborn, G.: The Ideology of Power and the Power of Ideology. NLB Publications, London (1980)

[R7.28] Thompson, K.: Beliefs and Ideology. Tavistock Publication, New York (1986)

[R7.29] Ziman, J.: The Problem of 'Problem Choice'. Minerva 25, 92–105 (1987)

[R7.30] Ziman, J.: Public Knowledge: An Essay Concerning the Social Dimension of Science. Cambridge University Press, Cambridge (1968)

[R7.31] Zuckerman, H.: Theory Choice and Problem Choice in Science. Sociological Inquiry 48, 65–95 (1978)

R8. Information, Thought and Knowledge

[R8.1] Aczel, J., Daroczy, Z.: On Measures of Information and their Characterizations. Academic Press, New York (1975)

[R8.2] Aczel, J., Forte, B.: A System of Axioms for the Measure of the Uncertainty. Notices, Amer. Math. Soc. 17, 202 (1970)

[R8.3] Anderson, J.R.: The Architecture of Cognition. Harvard University Press, Cambridge (1983)

[R8.4] Angelov, S., Georgiev, D.: The Problem of Human Being in Contemporary Scientic Knowledge. Soviet Studies in Philosophy, 49–66 (Summer 1974)

[R8.5] Ash, R.: Information Theory. John Wiley and Sons, New York (1965)

[R8.6] Barlas, Y., Carpenter, S.: Philosophical Roots of Model Validation: Two Paradigms. System Dynamic Review 6, 148–166 (1990)

[R8.7] Barrett, T.W.: Quantum Statistical Foundations for Structural Information Theory and Communication Theory. In: Lakshmikantham, V. (ed.) Nonlinear Systems and Applications, pp. 389–407. Academic Press, New York (1977)

[R8.8] Bergin, J.: Common Knowledge with Monotone Statistics. Econometrica 69, 1315–1332 (2001)

[R8.9] Bestougeff, H., Ligozat, G.: Logical Tools for Temporal Knowledge Representation. Ellis Horwood, New York (1992)

[R8.10] Brillouin, L.: Science and information Theory. Academic Press, New York (1962)

[R8.11] Bruner, J.S., et al.: A Study of Thinking. Wiley, New York (1956)

[R8.12] Brunner, K., Meltzer, A.H. (eds.): Three Aspects of Policy and Policy Making: Knowledge, Data and Institutions. Carnegie-Rochester Conference Series, vol. 10. North-Holland, Amsterdam (1979)

[R8.13] Burks, A.W.: Chance, Cause, Reason: An Inquiry into the Nature of Scientific Evidence. University of Chicago Press, Chicago (1977)

[R8.14] Calvert, R.: Models of Imperfect Information in Politics. Hardwood Academic Publishers, New York (1986)

[R8.15] Cornforth, M.: The Theory of Knowledge. International Pub., New York (1972)

[R8.16] Coombs, C.H.: A Theory of Data. Wiley, New York (1964)

[R8.17] Crawshay-Willims: Methods and Criteria of Reasoning. Routledge and Kegan Paul, London (1957)

[R8.18] Dretske, F.I.: Knowledge and the Flow of Information. MIT Press, Cambridge (1981)

[R8.19] Dreyfus, H.L.: A Framework for Misrepresenting Knowledge. In: Ringle, M. (ed.) Philosophical Perspectives in Artificial Intelligence. Humanities Press, Atlantic Highlands (1979)

[R8.20] Fagin, R., Halpenn, J.Y.: Reasoning About Knowledge and Probability. In: Vardi, M. (ed.) Proceedings of the Second Conference of Theoretical Aspects of Reasoning about Knowledge, pp. 277–293. Morgan Kaufmann, Asiloman (1988)

[R8.21] Fagin, R., et al.: Reasoning About Knowledge. MIT Press, Cambridge (1995)

[R8.22] Geanakoplos, J.: Common Knowledge. In: Moses, Y. (ed.) Proceedings of the Fourth Conference of Theoretical Aspects of Reasoning about Knowledge. Morgan Kaufmann, San Mateo (1992)

[R8.23] Geanakoplos, J.: Common Knowledge. Journal of Economic Perspectives 6, 53–82 (1992)

[R8.24] George, F.H.: Models of Thinking. Allen and Unwin, London (1970)

[R8.25] George, F.H.: Epistemology and the problem of perception. Mind 66, 491–506 (1957)

[R8.26] Harwood, E.C.: Reconstruction of Economics. American Institute for Economic Research, Great Barrington (1955)

[R8.27] Hintikka, J.: Knowledge and Belief. Cornell University Press, Ithaca (1962)

[R8.28] Hirshleifer, J.: The Private and Social Value of Information and Reward to inventive activity. American Economic Review 61, 561–574 (1971)

[R8.29] Hirshleifer, J., Riley, J.: The Analytics of Uncertainty and Information: An expository Survey. Journal of Economic Literature 17, 1375–1421 (1979)

[R8.30] Hirshleifer, J., Riley, J.: The Economics of Uncertainty and Information. Cambridge University Press, Cambridge (1992)

[R8.31] Kapitsa, P.L.: The Influence of Scientific Ideas on Society. Soviet Studies in Philosophy, 52–71 (Fall 1979)

[R8.32] Kedrov, B.M.: The Road to Truth. Soviet Studies in Philosophy 4, 3–53 (1965)

[R8.33] Klatzky, R.L.: Human Memory: Structure and Processes. W. H. Freeman Pub., San Francisco (1975)

[R8.34] Koopmans, T.C.: Three Essays on the State of Economic Science. McGraw-Hill, New York (1957)

[R8.35] Kreps, D., Wilson, R.: Reputation and Imperfect Information. Journal of Economic Theory 27, 253–279 (1982)

[R8.36] Kubát, L., Zeman, J. (eds.): Entropy and Information. Elsevier, Amsterdam (1975)

[R8.37] Kurcz, G., Shugar, W., et al. (eds.): Knowledge and Language. North-Holland, Amsterdam (1986)

[R8.38] Lakemeyer, G., Nobel, B. (eds.): Foundations of Knowledge Representation and Reasoning. Springer, Berlin (1994)

[R8.39] Lektorskii, V.A.: Principles involved in the Reproduction of Objective in Knowledge. Soviet Studies in Philosophy 4(4), 11–21 (1967)

[R8.40] Levi, I.: The Enterprise of Knowledge. MIT Press, Cambridge (1980)

[R8.41] Levi, I.: Ignorance, Probability and Rational Choice. Synthese 53, 387–417 (1982)

[R8.42] Levi, I.: Four Types of Ignorance. Social Science 44, 745–756

[R8.43] Levine, D., Aparicio IV, M. (eds.): Neural Networks for Knowledge Representation and Inference. Lawrence Erlbaum Associates Publishers, Hillsdale (1994)

[R8.44] Marschak, J.: Economic Information, Decision and Prediction: Selected Essays, Boston, Mass., Dordrecnt-Holland, vol. II, Part II (1974)

[R8.45] McDermott, J.: Representing Knowledge in Intelligent Systems. In: Ringle, M. (ed.) Philosophical Perspectives in Artificial Intelligence, pp. 110–123. Humanities Press, Atlantic Highlands (1979)

[R8.46] Menges, G. (ed.): Information, Inference and Decision. D. Reidel Pub., Dordrecht (1974)

[R8.47] Masuch, M., Pólos, L. (eds.): Knowledge Representation and Reasoning Under Uncertainty. Springer, New York (1994)

[R8.48] Moses, Y. (ed.): Proceedings of the Fourth Conference of Theoretical Aspects of Reasoning about Knowledge. Morgan Kaufmann, San Mateo (1992)

[R8.49] Nielsen, L.T., et al.: Common Knowledge of Aggregation Expectations. Econometrica 58, 1235–1239 (1990)

[R8.50] Newell, A.: Unified Theories of Cognition. Harvard University Press, Cambridge (1990)

[R8.51] Newell, A.: Human Problem Solving. Prentice-Hall, Englewood Cliff (1972)

[R8.52] Ogden, G.K., Richards, I.A.: The Meaning of Meaning. Harcourt-Brace Jovanovich, New York (1923)

[R8.53] Planck, M.: Scientific Autobiography and Other Papers. Greenwood, Westport (1968)

[R8.54] Pollock, J.: Knowledge and Justification. Princeton University Press, Princeton (1974)

[R8.55] Polanyi, M.: Personal Knowledge. Routledge and Kegan Paul, London (1958)

[R8.56] Popper, K.R.: Objective Knowledge. Macmillan, London (1949)

[R8.57] Price, H.H.: Thinking and Experience. Hutchinson, London (1953)

[R8.58] Putman, H.: Reason, Truth and History. Cambridge University Press, Cambridge (1981)

[R8.59] Putman, H.: Realism and Reason. Cambridge University Press, Cambridge (1983)

[R8.60] Putman, H.: The Many Faces of Realism. Open Court Publishing Co., La Salle (1987)

[R8.61] Rothschild, K.W.: Models of Market Organization with Imperfect Information: A Survey. Journal of Political Economy 81, 1283–1308 (1973)

[R8.62] Russell, B.: Human Knowledge, its Scope and Limits. Allen and Unwin, London (1948)

[R8.63] Russell, B.: Our Knowledge of the External World. Norton, New York (1929)

[R8.64] Samet, D.: Ignoring Ignorance and Agreeing to Disagree. Journal of Economic Theory 52, 190–207 (1990)

[R8.65] Schroder, H.M., Suedfeld, P. (eds.): Personality Theory and Information Processing. Ronald Pub., New York (1971)

[R8.66] Searle, J.: Minds, Brains and Science. Harvard University Press, Cambridge (1985)

[R8.67] Sen, A.K.: On Weights and Measures: Information Constraints in Social Welfare Analysis. Econometrica 45, 1539–1572 (1977)

[R8.68] Shin, H.: Logical Structure of Common Knowledge. Journal of Economic Theory 60, 1–13 (1993)

[R8.69] Simon, H.A.: Models of Thought. Yale University Press, New Haven (1979)

[R8.70] Smithson, M.: Ignorance and Uncertainty, Emerging Paradigms. Springer, New York (1989)

[R8.71] Sowa, J.F.: Knowledge Representation: Logical, Philosophical, and Computational Foundations. Brooks Pub., Pacific Grove (2000)

[R8.72] Stigler, G.J.: The Economics of Information. Journal of Political Economy 69, 213–225 (1961)

[R8.73] Tiukhtin, V.S.: How Reality Can be Reflected in Cognition: Reflection as a Property of All Matter. Soviet Studies in Philosophy 3(1), 3–12 (1964)

[R8.74] Tsypkin, Y.Z.: Foundations of the Theory of Learning Systems. Academic Press, New York (1973)

[R8.75] Ursul, A.D.: The Problem of the Objectivity of Information. In: Kubát, L., Zeman, J. (eds.) Entropy and Information, pp. 187–230. Elsevier, Amsterdam (1975)

[R8.76] Vardi, M. (ed.): Proceedings of Second Conference on Theoretical Aspects of Reasoning about Knowledge. Morgan Kaufmann, Asiloman (1988)

[R8.77] Vazquez, M., et al.: Knowledge and Reality: Some Conceptual Issues in System Dynamics Modeling. Systems Dynamics Review 12, 21–37 (1996)

[R8.78] Zadeh, L.A.: A Theory of Commonsense Knowledge. In: Skala, H.J., et al. (eds.) Aspects of Vagueness, pp. 257–295. D. Reidel Co., Dordrecht (1984)

[R8.79] Zadeh, L.A.: The Concept of Linguistic Variable and its Application to Approximate reasoning. Information Science 8, 199–249 (1975); also in 9, 40–80

R9. Language and the Knowledge-Production Process

[R9.1]   Aho, A.V.: Indexed Grammar - An Extension of Context-Free Grammars. Journal of the Association for Computing Machinery 15, 647–671 (1968)

[R9.2]   Black, M. (ed.): The Importance of Language. Prentice-Hall, Englewood Cliffs (1962)

[R9.3]   Buchler, J.: Metaphysics of Natural Complexes. Columbia University Press, New York (1966)

[R9.4]   Carnap, R.: Meaning and Necessity: A Study in Semantics and Modal Logic. University of Chicago Press, Chicago (1956)

[R9.5]   Chomsky, N.: Linguistics and Philosophy. In: Hook, S. (ed.) Language and Philosophy, pp. 51–94. New York University Press, New York (1968)

[R9.6]   Chomsky, N.: Language and Mind. Harcourt Brace Jovanovich, New York (1972)

[R9.7]   Cooper, W.S.: Foundations of Logico-Linguistics: A Unified Theory of Information, Language and Logic. D. Reidel, Dordrecht (1978)

[R9.8]   Cresswell, M.J.: Logics and Languages. Methuen Pub., London (1973)

[R9.9]   Dilman, I.: Studies in Language and Reason. Barnes and Nobles, Books, Totowa, N.J (1981)

[R9.10] Fodor, J.A.: The Language and Thought. Thomas Y. Crowell Co., New York (1975)

[R9.11] Ginsbury, S.: Algebraic and Automata – Theoretical properties of Formal Languages. North-Holland, New York (1975)

[R9.12] Givon, T.: On Understanding Grammar. Academic Press, New York (1979)

[R9.13] Gorsky, D.R.: Definition. Progress Publishers, Moscow (1974)

[R9.14] Greibach, S.A.: An Infinite Hierarchy of Contex-Free Languages. Journal of Association for Computing Machinery 16, 91–106 (1969)

[R9.15] Hintikka, J.: The Game of Language. D. Reidel Pub., Dordrecht (1983)

[R9.16] Johnson-Lair, P.N.: Mental Models: Toward Cognitive Science of Language, Inference and Consciousness. Harvard University Pres, Cambridge (1983)

[R9.17] Kandel, A.: Codes Over Languages. IEEE Transactions on Systems Man and Cybernetics 4, 135–138 (1975)

[R9.18] Keenan, E.L., Faltz, L.M.: Boolean Semantics for Natural Languages. D. Reidel Pub., Dordrecht (1985)

[R9.19] Lakoff, G.: Linguistics and Natural Logic. Synthese 22, 151–271 (1970)

[R9.20] Lee, E.T., et al.: Notes On Fuzzy Languages. Information Science 1, 421–434 (1969)

[R9.21] Mackey, A., Merrill, D. (eds.): Issues in the Philosophy of Language. Yale University Press, New Haven (1976)

[R9.22] Nagel, T.: Linguistics and Epistemology. In: Hook, S. (ed.) Language and Philosophy, pp. 180–184. New York University Press, New York (1969)

[R9.23] Pike, K.: Language in Relation to a Unified Theory of Structure of Human Behavior. Mouton Pub., The Hague (1969)

[R9.24] Quine, W.V.O.: Word and object. MIT Press, Cambridge (1960)

[R9.25] Russell, B.: An Inquiry into Meaning and Truth. Penguin Books (1970)

[R9.26] Salomaa, A.: Formal Languages. Academic Press, New York (1978)

[R9.27] Tamura, S., et al.: learning of Formal Language. IEEE Trans. Syst. Man. Cybernetics, SMC-3, 98–102 (1973)

[R9.28] Tarski, A.: Logic, Semantics and Matamathematics. Clarendon Press, Oxford (1956)

[R9.29] Whorf, B.L. (ed.): Language, Thought and Reality. Humanities Press, Language (1956)

[R9.30] Winogrand, T.: Understanding Natural Language. Cognitive Psychology 3, 1–191 (1972)

R10. Probabilistic Concepts and Reasoning

[R10.1] Anscombe, F., Aumann, R.J.: A Definition of Subjective Probability. Annals of Mathematical Statistics 34, 199–205 (1963)

[R10.2] Billingsley, P.: Probability and Measure. John Wiley and Sons, Chichester (1979)

[R10.3] Boolos, G.S., Jeffrey, R.C.: Computability and Logic. Cambridge University Press, New York (1989)

[R10.4] Carnap, R.: Logical Foundation of Probability. Routledge and Kegan Paul Ltd., London (1950)

[R10.5] Cohen, L.J.: The Probable and Provable. Clarendon Press, Oxford (1977)

[R10.6] de Finetti, B.: Probabilities of Probabilities a Real Problem or Misunderstanding? In: Aykac, A., et al. (eds.) New Developments in the Applications of Bayesian Methods, Amsterdam, pp. 1–10 (1977)

[R10.7] Dempster, A.P.: Upper and Lower Probabilities Induced by Multivalued Mapping. Annals of Math Statistics 38, 325–339 (1967)

[R10.8] Domotor, Z.: Higher Order Probabilities. Philosophical Studies 40, 31–46 (1981)

[R10.9] Doob, J.L.: Stochastic Processes. John Wiley and Sons, New York (1990)

[R10.10] Fellner, W.: Distortion of Subjective Probabilities as a Reaction to Uncertainty. Quarterly Journal of Economics 75, 670–689 (1961)

[R10.11] Fishburn, P.C.: The Axioms of Subjective Probability. Statistical Sciences 1(3), 335–358 (1986)

[R10.12] Fishburn, P.C.: Decision and Value. John Wiley and Sons, New York (1964)

[R10.13] Gaifman, C.: A Theory of Higher Order Probabilities. In: Halpern, J.Y. (ed.) Theoretical Aspects of Reasoning about Knowledge, pp. 275–292. Morgan Kaufman, Los Altos (1986)

[R10.14] George, F.H.: Logical Networks and Probability. Bulletin of Mathematical Biophysics 19, 187–199 (1957)

[R10.15] Good, I.J.: Probability and the Weighing of Evidence. Charles Griffin and Co. Ltd, London (1950)

[R10.16] Good, I.J.: Subjective Probability as the Measure of Non-measurable Set. In: Nagel, E., et al. (eds.) Logic, Methodology, and the Philosophy of Science, pp. 319–329. Stanford University Press, Stanford (1962)

[R10.17] Good, I.J.: Good Thinking: The Foundations of Probability and Applications. University of Minnesota Press, Minneapolis (1983)

[R10.18] Goutsias, J., et al. (eds.): Random Sets: Theory and Applications. Springer, New York (1997)

[R10.19] Hacking, I.: The emergence of Probability. Cambridge University Press, London (1975)

[R10.20] Harsanyi, J.C.: Acceptance of Empirical Statements: A Bayesian Theory without Cognitive Utilities. Theory and Decision 18, 1–30 (1985)

[R10.21] Ulrich, H., Klement, E.P.: Plausibility Measures: A General Framework for Possibility and Fuzzy Probability Measures. In: Skala, H.J., et al. (eds.) Aspects of Vagueness, pp. 31–50. D Reidel Pub. Co., Dordrecht (1984)

[R10.22] Holmos, P.R.: Measure Theory. Van Nostrand, New York (1950)

[R10.23] Hoover, D.N.: Probabilistic Logic. Annals of Mathematical Logic 14, 287–313 (1978)

[R10.24] Jeffery, R.: The Present Position in Probability Theory. British Journal for the Philosophy of Science 5, 275–280 (1955)

[R10.25] Kahneman, D., Tversky, A.: Sujective Probability: A Judgment of representativeness. Cognitive Psychology 3, 430–454 (1972)

[R10.26] Keynes, J.M.: A treatise on Probability. MacMillan and Co., London (1921)

[R10.27] Kolmogrov, A.N.: Foundation of the Theory of Probability. Chelsea Pub. Co., New York (1956)

[R10.28] Koopman, B.O.: The Axioms and Algebra of Intuitive Probability. Annals of Mathematics 41, 269–292 (1940)

[R10.29] Kraft, C., et al.: Intuitive Probability on Finite Sets. Annals of Mathematical Statistics 30, 408–419 (1959)

[R10.30] Kullback, S., Leibler, R.A.: Information and Sufficiency. Annals of Math. Statistics 22, 79–86 (1951)

[R10.31] Kyburg, H.E.: Probability and the Logic of Rational Belief. Wesleyan University Press, Middleton (1961)

[R10.32] Kyburg, H.E., Smokler, H.E.: Studies in SubjectiveProbability. Wiley, Chichester (1964)

[R10.33] Laha, R., Rohatgi, V.K.: Probability Theory. John Wiley and Sons, New York (1979)

[R10.34] Laplace, P.S.: A Philosophical Essay on Probabilities. Constable and Co., London (1951)

[R10.35] Matheron, G.: Random Sets and Integral Geometry. John Wiley and Sons, New York (1975)

[R10.36] Marschak, J.: Personal Probabilities of Probabilities. Theory and Decision 6, 121–153 (1975)

[R10.37] Nagel, E.: Principles of the Theory of Probability. In: Neurath, O., et al. (eds.) International Encyclopedia of Unified Science, vol. 1 -10, pp. 343–422. University of Chicago Press, Chicago (1955)

[R10.38] Nilsson, N.J.: Probabilistic Logic. Artificial Intelligence 28, 71–87 (1986)

[R10.39] Parrat, L.G.: Probability and Experimental Errors in Science. John Wiley and Sons, New York (1961)

[R10.40] Patty Wayne, C.: Foundations of Topology. PWS Pub. Co., Boston (1993)

[R10.41] Parthasarath, K.R.: Probability Measure on Metric Spaces. Academic Press, New York (1967)

[R10.42] Ruspini, E.: Epistemic Logics, Probability and Calculus of Evidence. In: Proceedings of 10th International Joint Conference on AI (IJCAI 1987), Milan, pp. 924–931 (1987)

[R10.43] Savage, L.J.: The Foundations of Statistics. John Wiley and Sons, New York (1954)

[R10.44] Schneeweiss, H.: Probability and Utility – Dual Concepts in Decision Theory I. In: Menges (ed.) Information, Inference and Decision. Reidel, Dordrecht (1974)

[R10.45] Shafer, G.: A Mathematical Theory of Evidence. Princeton University Press, Princeton (1976)

[R10.46] Shafer, G.: Constructive Probability. Synthese 48, 1–60 (1997)

[R10.47] Shannon, C.E., Weaver, W.: The Mathematical Theory of Communication. University of Illinois Press, Urbana (1949)

[R10.48] Tiller, P., Green, E.D. (eds.): Probability and Inference in the Law of Evidence: The Uses and Limits of Bayesianism. Kluwer Academic Publishers, Dordrecht (1988)

[R10.49] von Mises, R.: Probability, Statistics and Truth. Dover Pub., New York (1981)

[R10.50] Wagon, S.: The Banach-Tarski Paradox. Cambridge University Press, Cambridge (1985)

R11. Optimality, Classical Exactness and Equilibrium in Knowledge Systems

[R11.1] Agassi, J., Jarvie, I.C. (eds.): Rationality: The Critical View. M. Nijhoff Pub., Boston (1987)

[R11.2] Amsterdamski, S.: Between History and Method: Disputes about the Rationality of Science. Kluwer Academic Pub., Dordrecht (1992)

[R11.3] Anderson, G.: Rationality in Science and Politics. D. Reid Pub. Co., Dordrecht (1984)

[R11.4] Arrow, K.J.: Rationality of self and Others in an Economic System. Journal of Business 59, S385–S399 (1986)

[R11.5] Baumol, W., Quant, R.: Rules of Thumb and Optimally Imperfect Decisions. American Economic Review 54, 23–46 (1964)

[R11.6] Benn, S.I., Mortimore, G.W. (eds.): Rationality and the Social Sciences: Contributions to the Philosophy and Methodology of the Social Sciences. Routledge and Kegan Paul, London (1976)

[R11.7] Bicchieri, C.: Rationality and Coordination. Cambridge University Press, New York (1993)

[R11.8] Biderman, S., Scharfstein, B.-A.: Rationality in a question: East and Western View of Rationality. E.J. Brill Pub., New York (1989)

[R11.9] Black, F.: Exploring General Equilibrium. MIT Press, Cambridge (1995)

[R11.10] Boland, L.A.: On the Futility of Criticizing the Neoclassical Maximization Hypothesis. The American Economic Review 71, 1031–1036 (1981)

[R11.11] Border, K.C.: Fixed Point Theorems with Applications to Economics and Game Theory. Cambridge University Press, Cambridge (1985)

[R11.12] Bowman, E.H.: Consistency and Optimality in Managerial Decision Making. Management Science 9, 310–321 (1963)

[R11.13] Brubaker, R.: The Limits of Rationality: An Essay on the Social and Moral Thought of Max Weber. Allen and Unwin, London (1984)

[R11.14] Churchman, C.W.: Prediction and Optimal Decision. Pretice-Hall, Englewood Cliffs (1961)

[R11.15] Cohen, L.J.: Can Human Irrationality be Experimentally Demonstrated? Behavioral and Brain Science 4, 317–370 (1981)

[R11.16] Cohen, M., Jaffray, J.-Y.: Is Savage's Independence Axiom a Universal Rationality Principle? Behavioral Science 33, 38–47

[R11.17] Cornwall, R.R.: Introduction to the use of General Equilibrium Analysis. North-Holland, New York (1984)

[R11.18] De Sousa, R.: The Rationality of Motion. MIT Press, Cambridge (1987)

[R11.19] Dompere, K.K.: Fuzziness, Rationality, Optimality and Equilibrium in Decision and Economic Theories. In: Lodwick, W.A., Kacprzyk, J. (eds.) Fuzzy Optimization. STUDFUZZ, vol. 254, pp. 3–32. Springer, Heidelberg (2010)

[R11.20] Dompere, K.K.: On Epistemology and Decision-Choice Rationality. In: Trappl, R. (ed.) Cybernetics and System Research, pp. 219–228. North-Holland, New York (1982)

[R11.21] Dompere, K.K.: Epistemic Foundations of Fuzziness: Unified Theories of Decision-Choice Processes. STUDFUZZ, vol. 236. Springer, New York (2009)

[R11.22] Dompere, K.K.: Fuzziness and Approximate Reasoning: Epistemics on Uncertainty, Expectations and Risk in Rational Behavior. STUDFUZZ, vol. 237. Springer, New York (2009)

[R11.23] Elster, J.: Ulysses and the Sirens: Studies in Rationality and Irrationality. Cambridge University Press, New York (1979)

[R11.24] Elster, J.: Studies in the Subversion of Rationality. Cambridge University Press, New York (1983)

[R11.25] Ernst, G.C., et al.: Principles of Structural Equilibrium, a Study of Equilibrium Conditions by Graphic, Force Moment and Virtual Displacement (virtual work). University of Nebraska Press, Lincoln Na (1962)

[R11.26] Fischer, R.B., Peters, G.: Chemical Equilibrium. Saunders Pub., Philadelphia (1970)

[R11.27] Fisher, F.M.: Disequilibrium Foundations of Equilibrium Economics. Cambridge University Press, New York (1983)

[R11.28] Fourgeaud, C., Gouriéroux, C.: Learning Procedures and Convergence to Rationality. Econometrica 54, 845–868 (1986)

[R11.29] Freeman, A., Carchedi, G. (eds.): Marx and Non-Equilibrium Economics. Edward Elgar, Cheltenham (1996)

[R11.30] Newton, G., Hare, P.H. (eds.): Naturalism and Rationality. Prometheus Books, Buffalo (1986)

[R11.31] Ginsburgh, V.: Activity Analysis and General Equilibrium Modelling. North-Holland, New York (1981)

[R11.32] Hahn, F.: Equilibrium and Macroeconomics. MIT Press, Cambridge (1984)

[R11.33] Hansen, B.: A Survey of General Equilibrium Systems. McGraw-Hill, New York (1970)

[R11.34] Hilpinen, R. (ed.), Rationality in Science: Studies in the Foundations of Science and Ethics. D. Reidel, Dordrecht (1980)

[R11.35] Hogarth, R.M., Reder, M.W. (eds.): Rational Choice: The Contrast between Economics and Psychology. University of Chicago Press, Chicago (1986)

[R11.36] Hollis, M., Lukes, S. (eds.): Rationality and Relativism. MIT Press, Cambridge (1982)

[R11.37] Martin, H., Nell, E.J.: Rational Economic Man: A Philosophical Critique of Neo-classical Economics. Cambridge University Press, New York (1975)

[R11.38] Howard, N.: Paradoxes of Rationality. MIT Press, Cambridge (1973)

[R11.39] Hurwicz, L.: Optimality Criteria for Decision Making Under Ignorance, Mimeographed, Cowles Commission Discussion Paper, Statistics 370 (1951)

[R11.40] Istratescu, V.I.: Fixed Point Theory: Introduction. D. Reidel Pub. Co, Dordrecht (1981)

[R11.41]   Keita, L.D.: Science, Rationality and Neoclassical Economics. University of Delaware Press, Newark (1992)

[R11.42]   Kirman, A. (ed.): Elements of General Equilibrium Analysis. Blackwell, Malden (1998)

[R11.43]   Kirman, A., Salmon, M. (eds.): Learning and Rationality in Economics. Basil Blackwell, Oxford (1993)

[R11.44]   Kornia, J.: Anti-Equilibrium. North-Holland, Amsterdam (1971)

[R11.45]   Kramer, G.H.: A Dynamic Model of Political Equilibrium. Journal of Economic Theory 16, 310–334 (1977)

[R11.46]   Kramer, G.: On a Class of Equilibrium Conditions for Majority Rule. Econometrica 41, 285–297 (1973)

[R11.47]   Kuenne, R.E.: The Theory of General Economic Equilibrium. Princeton University Press, Princeton (1967)

[R11.48]   Marschak, J.: Actual Versus Consistent Decision Behavior. Behavioral Science 9, 103–110 (1964)

[R11.49]   McKenzie, L.W.: Classical General Equilibrium Theory. MIT Press, Cambridge (2002)

[R11.50]   McMullin, E. (ed.): Construction and Constraint: The Shaping of Scientific Rationality. University of Notre Dame Press, Notre Dame Ind. (1988)

[R11.51]   Jozef, M. (ed.): The Problem of Rationality in Science and its Philosophy: Popper vs. Polanyi (The Polish Conferences). Kluwer Academic Publishers, Boston (1995)

[R11.52]   Mongin, P.: Does Optimization Imply Rationality. Synthese 124, 73–111 (2000)

[R11.53]   Negishi, T.: Microeconomic Foundations of Keynesian Macroeconomics. North-Holland, Amsterdam (1979)

[R11.54]   Newton-Smith, W.H.: The Rationality of Science. Routledge and Kegan Paul, Boston (1981)

[R11.55]   Page, S.E.: Two Measures of Difficulty. Economic Theory 8, 321–346 (1996)

[R11.56]   Parfit, D.: Personal Identity and Rationality. Synthese 53, 227–241 (1982)

[R11.57]   Pitt, J.C., Pera, M. (eds.): Rational Changes in Science: Essays on Scientific Reasoning. D. Reidel, Dordrecht (1987)

[R11.58]   Plott, C.R.: A Notion of Equilibrium and Its Possibility under Majority Rule. Amer. Econ. Rev. 57, 787–806 (1967)

[R11.59]   Preston, C.J.: Random Fields. Springer, Berlin (1976)

[R11.60]   Quirk, J., Saposnik, R.: Introduction to General Equilibrium Theory and Welfare Economics. McGraw-Hill, New York (1968)

[R11.61]   Radner, R.: Competitive Equilibrium Under Uncertainty. Econometrica 36, 31–58 (1968)

[R11.62]   Rapart, A.: Escape from Paradox. Scientific American, 50–56 (July 1967)

[R11.63]   Rescher, N. (ed.): Reason and Rationality in Natural Science: A Group of Essays. University Press of America, Lanham (1985)

[R11.64]   Sen, A.: Rational Behavior. In: Eatwell, J., et al. (eds.) Utility and Probability, New York, Norton, pp. 198–216 (1990)

[R11.65]   Shackle, G.: Epistemics and Economics. Cambridge University Press, Cambridge (1973)

[R11.66]   Simon, H.A.: Rational Choice and the Structure of Environment. Psychological Review 63, 129–138 (1956)

[R11.67]   Simon, H.A.: Rationality as Process and as Product of Thought. American Economic Review 68, 1–16 (1978)

[R11.68] Simon, H.A.: A Behavioral Model of Rational Choice. Quarterly Journal of Economics 69, 99–118 (1955)

[R11.69] Simon, H.A.: Models of Man: Social and Rational. Wily, New York (1957)

[R11.70] Smart, D.R.: Fixed Point Theorems. Cambridge University Press, Cambridge (1980)

[R11.71] Stambaugh, J.: The Real is not the Rational. State University of New York Press, Albany (1986)

[R11.72] Tamny, M., Irani, K.D. (eds.): Rationality in Thought and Action. Greenwood Press, New York (1986)

[R11.73] Turski, W.G.: Toward a Rationality of Emotions; Essay in the Philosophy of Mind. Ohio University Press, Athens (1994)

[R11.74] Torr, C.: Equilibrium, Expectations, and Information: A Study of General Theory and Modern Classical Economics. Westview Press, Boulder Colo (1988)

[R11.75] Valentinuzzi, M.: The Organs of Equilibrium and Orientation as a Control System. Hardwood Academic Pub., New York (1980)

[R11.76] Walsh, V.C., Gram, H.: Classical and Neoclassical Theories of General Equilibrium: Historical Origins and Mathematical Structure. Oxford University Press, New York (1980)

[R11.77] Roy, W.E.: General Equilibrium Analysis: Studies in Appraisal. Cambridge University Press, Cambridge (1985)

[R11.78] Roy, W.E.: Microfoundations: The Compatibility of Microeconomics and Macroeconomics. Cambridge University Press, Cambridge (1980)

[R11.79] Whittle, P.: Systems in Stochastic Equilibrium. Wiley, New York (1986)

R12. Possible Worlds and the Knowledge Production Process

[R12.1] Adams, R.M.: Theories of Actuality. Noûs 8, 211–231 (1974)

[R12.2] Allen, S. (ed.): Possible Worlds in Humanities, Arts and Sciences, Proceedings of Nobel Symposium, vol. 65. Walter de Gruyter Pub., New York (1989)

[R12.3] Armstrong, D.M.: A Combinatorial Theory of Possibility. Cambridge University Press (1989)

[R12.4] Armstrong, D.M.: A World of States of Affairs. Cambridge University Press, Cambridge (1997)

[R12.5] Bell, J.S.: Six Possible Worlds of Quantum Mechanics. In: Allen, S. (ed.) Proceedings of Nobel Symposium Possible Worlds in Humanities, Arts and Sciences, vol. 65, pp. 359–373.... Walter de Gruyter Pub., New York (1989)

[R12.6] Bigelow, J.: Possible Worlds Foundations for Probability. Journal of Philosophical Logic 5, 299–320 (1976)

[R12.7] Bradley, R., Swartz, N.: Possible World: An Introduction to Logic and its Philosophy. Bail Blackwell, Oxford (1997)

[R12.8] Castañeda, H.-N.: Thinking and the Structure of the World. Philosophia 4, 3–40 (1974)

[R12.9] Chihara, C.S.: The Worlds of Possibility: Modal Realism and the Semantics of Modal Logic, Clarendon (1998)

[R12.10] Chisholm, R.: Identity through Possible Worlds: Some Questions. Noûs 1, 1–8 (1967); reprinted in Loux, The Possible and the Actual

[R12.11] Divers, J.: Possible Worlds. Routledge, London (2002)

[R12.12] Forrest, P.: Occam's Razor and Possible Worlds. Monist 65, 456–464 (1982)

[R12.13] Forrest, P., Armstrong, D.M.: An Argument Against David Lewis' Theory of Possible Worlds. Australasian Journal of Philosophy 62, 164–168 (1984)

[R12.14] Grim, P.: There is No Set of All Truths. Analysis 46, 186–191 (1986)

[R12.15] Heller, M.: Five Layers of Interpretation for Possible Worlds. Philosophical Studies 90, 205–214 (1998)

[R12.16] Herrick, P.: The Many Worlds of Logic. Oxford University Press, Oxford (1999)

[R12.17] Krips, H.: Irreducible Probabilities and Indeterminism. Journal of Philosophical Logic 18, 155–172 (1989)

[R12.18] Kuhn, T.S.: Possible Worlds in History of Science. In: Allen, S. (ed.) Proceedings of Nobel Symposium on Possible Worlds in Humanities, Arts and Sciences, vol. 65, pp. 9–41. Walter de Gruyter Pub., New York (1989)

[R12.19] Kuratowski, K., Mostowski, A.: Set Theory: With an Introduction to Descriptive Set Theory. North-Holland, New York (1976)

[R12.20] Lewis, D.: On the Plurality of Worlds. Basil Blackwell, Oxford (1986)

[R12.21] Loux, M.J. (ed.): The Possible and the Actual: Readings in the Metaphysics of Modality. Cornell University Press, Ithaca (1979)

[R12.22] Parsons, T.: Nonexistent Objects. Yale University Press, New Haven (1980)

[R12.23] Perry, J.: From Worlds to Situations. Journal of Philosophical Logic 15, 83–107 (1986)

[R12.24] Rescher, N., Brandom, R.: The Logic of Inconsistency: A Study in Non-Standard Possible-World Semantics And Ontology, Rowman and Littlefield (1979)

[R12.25] Skyrms, B.: Possible Worlds, Physics and Metaphysics. Philosophical Studies 30, 323–332 (1976)

[R12.26] Stalmaker, R.C.: Possible World. Noûs 10, 65–75 (1976)

[R12.27] Quine, W.V.O.: Word and Object. M.I.T. Press (1960)

[R12.28] Quine, W.V.O.: Ontological Relativity. Journal of Philosophy 65, pp. 185–212 (1968)

R13. Rationality, Information, Games, Conflicts and Exact Reasoning

[R13.1] Aumann, R.: Correlated Equilibrium as an Expression of Bayesian Rationality. Econometrica 55, 1–18 (1987)

[R13.2] Border, K.: Fixed Point Theorems with Applications to Economics and Game Theory. Cambridge University Press, Cambridge (1985)

[R13.3] Brandenburger, A.: Knowledge and Equilibrium Games. Journal of Economic Perspectives 6, 83–102 (1992)

[R13.4] Campbell, R., Sowden, L.: Paradoxes of Rationality and Cooperation: Prisoner's Dilemma and Newcomb's Problem. University of British Columbia Press, Vancouver (1985)

[R13.5] Crawford, V., Sobel, J.: Strategic Information Transmission. Econometrica 50, 1431–1452 (1982)

[R13.6] Scott, G., Humes, B.: Games, Information, and Politics: Applying Game Theoretic Models to Political Science. University of Michigan Press, Ann Arbor (1996)

[R13.7] Gjesdal, F.: Information and Incentives: The Agency Information Problem. Review of Economic Studies 49, 373–390 (1982)

[R13.8] Harsanyi, J.: Games with Incomplete Information Played by 'Bayesian' Players I: The Basic Model. Management Science 14, 159–182 (1967)

[R13.9]   Harsanyi, J.: Games with Incomplete Information Played by 'Bayesian' Players II: Bayesian Equilibrium Points. Management Science 14, 320–334 (1968)

[R13.10] Harsanyi, J.: Games with Incomplete Information Played by 'Bayesian' Players III: The Basic Probability Distribution of the Game. Management Science 14, 486–502 (1968)

[R13.11] Harsanyi, J.: Rational Behavior and Bargaining Equilibrium in Games and Social Situations. Cambridge University Press, New York (1977)

[R13.12] Haussmann, U.G.: A Stochastic Maximum Principle for Optimal Control of Diffusions. Longman, Essex (1986)

[R13.13] Istratescu, V.I.: Fixed Point Theory: An Introduction. Reidel Pub. Co., Dordrecht (1981)

[R13.14] Krasovskii, N.N., Subbotin, A.I.: Game-theoretical Control Problems. Springer, New York (1988)

[R13.15] Kuhn, H. (ed.): Classics in Game Theory. Princeton University Press, Princeton (1997)

[R13.16] Lagunov, V.N.: Introduction to Differential Games and Control Theory. Heldermann Verlag, Berlin (1985)

[R13.17] Luce, D.R., Raiffa, H.: Games and Decisions. John Wiley and Sons, New York (1957)

[R13.18] Maynard Smith, J.: Evolution and the Theory of Games. Cambridge University Press, Cambridge (1982)

[R13.19] Milgrom, P., Roberts, J.: Rationalizablility, Learning and Equilibrium in Games with Strategic Complementarities. Econometrica 58, 1255–1279 (1990)

[R13.20] Myerson, R.: Game Theory: Analysis of Conflict. Harvard University Press, Cambridge (1991)

[R13.21] Rapoport, A., Chammah, A.: Prisoner's Dilemma: A Study in Conflict and Cooperation. University of Michigan Press, Ann Arbor (1965)

[R13.22] Roth, A.E.: The Economist as Engineer: Game Theory, Experimentation, and Computation as Tools for Design Economics. Econometrica 70, 1341–1378 (2002)

[R13.23] Shubik, M.: Game Theory in the Social Sciences: Concepts and Solutions. MIT Press, Cambridge (1982)

[R13.24] Smart, D.R.: Fixes point Theorems. Cambridge University Press, Cambridge (1980)

[R13.25] Ulrich, H.: Fixed Point Theory of Parametrized Equivariant Maps. Springer, New York (1988)

[R13.26] Von Neumann, J., Morgenstern, O.: The Theory of Games in Economic Behavior. John Wiley and Sons, New York (1944)

R14. Rationality and Philosophy of Exact and Inexact Sciences in the Knoeledge Production

[R14.1]   Achinstein, P.: The Problem of Theoretical Terms. In: Brody, B.A. (ed.) Reading in the Philosophy of Science, Prentice Hall, Englewood Cliffs (1970)

[R14.2]   Amo Afer, A.G.: The Absence of Sensation and the Faculty of Sense in the Human Mind and Their Presence in our Organic and Living Body, Dissertation and Other essays 1727-1749, Halle Wittenberg, Jena, Martin Luther Universioty Translation (1968)

[R14.3] Beeson, M.J.: Foundations of Constructive Mathematics. Springer, Berlin (1985)

[R14.4] Benacerraf, P.: God, the Devil and Gödel. Monist 51, 9–32 (1967)

[R14.5] Benecerraf, P., Putnam, H. (eds.): Philosophy of Mathematics: Selected Readings. Cambridge University Press, Cambridge (1983)

[R14.6] Black, M.: The Nature of Mathematics. Adams and Co., Totowa (1965)

[R14.7] Blanche, R.: Contemporary Science and Rationalism. Oliver and Boyd, Edinburgh (1968)

[R14.8] Blanshard, B.: The Nature of Thought. Allen and Unwin, London (1939)

[R14.9] Bloomfield, L.: Linguistic Aspects of Science. In: Neurath, O., et al. (eds.) International Encyclopedia of Unified Science, vol. 1-10, pp. 219–277. University of Chicago Press, Chicago (1955)

[R14.10] Braithwaite, R.B.: Models in the empirical Sciences. In: Brody, B.A. (ed.) Reading in the Philosophy of Science, pp. 268–275. Prentice Hall, Englewood Cliffs (1970)

[R14.11] Braithwaite, R.B.: Scientific Explanation. Cambridge University Press, Cambridge (1955)

[R14.12] Brody, B.A. (ed.): Reading in the Philosophy of Science. Prentice Hall, Englewood Cliffs (1970)

[R14.13] Brody, B.A.: Confirmation and Explanation. In: Brody, B.A. (ed.) Reading in the Philosophy of Science, pp. 410–426. Prentice- Hall, Englewood Cliffs (1970)

[R14.14] Brouwer, L.E.J.: Intuitionism and Formalism. Bull. of American Math. Soc. 20, 81–96 (1913); Also in Benecerraf, P., Putnam, H. (eds.) Philosophy of Mathematics: Selected Readings, pp. 77–89. Cambridge University Press, Cambridge (1983)

[R14.15] Brouwer, L.E.J.: Consciousness, Philosophy, and Mathematics. In: Benecerraf, P., Putnam, H. (eds.) Philosophy of Mathematics: Selected Readings, pp. 90–96. Cambridge University Press, Cambridge (1983)

[R14.16] Brouwer, L.E.J.: Collected Works. In: Heyting, A. (ed.) Philosophy and Foundations of Mathematics, vol. 1. Elsevier, New York (1975)

[R14.17] Campbell, N.R.: What is Science? Dover, New York (1952)

[R14.18] Carnap, R.: Foundations of Logic and Mathematics. In: International Encyclopedia of Unified Science, pp. 143–211. Univ. of Chicago, Chicago (1939)

[R14.19] Carnap, R.: Statistical and Inductive Probability. In: Brody, B.A. (ed.) Reading in the Philosophy of Science, pp. 440–450. Prentice-Hall, Englewood Cliffs (1970)

[R14.20] Carnap, R.: On Inductive Logic. Philosophy of Science 12, 72–97 (1945)

[R14.21] Carnap, R.: The Two Concepts of Probability. Philosophy and Phenomenonological Review 5, 513–532 (1945)

[R14.22] Carnap, R.: The Methodological Character of Theoretical Concepts. In: Feigl, H., Scriven, M. (eds.) Minnesota Studies in the Philosophy of Science, vol. I, pp. 38–76 (1956)

[R14.23] Charles, D., Lennon, K. (eds.): Reduction, Explanation, and Realism. Oxford University Press, Oxford (1992)

[R14.24] Cohen, R.S., Wartofsky, M.W. (eds.): Methodological and Historical Essays in the Natural and Social Sciences. D. Reidel Publishing Co., Dordrecht (1974)

[R14.25] Church, A.: An Unsolvable Problem of Elementary Number Theory. American Journal of Mathematics 58, 345–363 (1936)

[R14.26] van Dalen, D. (ed.): Brouwer's Cambridge Lectures on Intuitionism. Cambridge University Press, Cambridge (1981)

[R14.27] Davidson, D.: Truth and Meaning: Inquiries into Truth and Interpretation. Oxford University Press, Oxford (1984)

[R14.28] Davis, M.: Computability and Unsolvability. McGraw-Hill, New York (1958)

[R14.29] Dompere, K.K.: Polyrhythmicity: Foundations of African Philosophy. Adonis and Abbey Pub., London (2006)

[R14.30] Dompere, K.K., Ejaz, M.: Epistemics of Development Economics: Toward a Methodological Critique and Unity. Greenwood Press, Westport (1995)

[R14.31] Dummett, M.: The Philosophical Basis of Intuitionistic Logic. In: Benecerraf, P., Putnam, H. (eds.) Philosophy of Mathematics: Selected Readings, pp. 97–129. Cambridge University Press, Cambridge (1983)

[R14.32] Feigl, H., Scriven, M. (eds.): Minnesota Studies in the Philosophy of Science, vol. I (1956)

[R14.33] Feigl, H., Scriven, M. (eds.): Minnesota Studies in the Philosophy of Science, vol. II (1958)

[R14.34] Frank, P.: Between Physics and Philosophy. Harvard University Press, Cambridge (1941)

[R14.35] Garfinkel, A.: Forms of Explanation: Structures of Inquiry in Social Science. Yale University Press, New Haven (1981)

[R14.36] Georgescu-Roegen, N.: Analytical Economics. Harvard University Press, Cambridge (1967)

[R14.37] George, F.H.: Philosophical Foundations of Cybernetics, Tunbridge Well, Great Britain (1979)

[R14.38] Gillam, B.: Geometrical Illusions. Scientific American, 102–111 (January 1980)

[R14.39] Gödel, K.: What is Cantor's Continuum Problem? In: Benecerraf, P., Putnam, H. (eds.) Philosophy of Mathematics: Selected Readings, pp. 470–486. Cambridge University Press, Cambridge (1983)

[R14.40] Gorsky, D.R.: Definition. Progress Publishers, Moscow (1974)

[R14.41] Gray, W., Rizzo, N.D. (eds.): Unity Through Diversity. Gordon and Breach, New York (1973)

[R14.42] Grattan-Guinness, I.: The Development of the Foundations of Mathematical Analysis From Euler to Riemann. MIT Press, Cambridge (1970)

[R14.43] Hart, W.D. (ed.): The Philosophy of Mathematics. Oxford University Press, Oxford (1996)

[R14.44] Hausman, D.M.: The Exact and Separate Science of Economics. Cambridge University Press, Cambridge (1992)

[R14.45] Helmer, O., Oppenheim, P.: A Syntactical Definition of Probability and Degree of confirmation. The Journal of Symbolic Logic 10, 25–60 (1945)

[R14.46] Helmer, O., Rescher, N.: On the Epistemology of the Inexact Sciences, P-1513. Rand Corporation, Santa Monica (1958)

[R14.47] Hempel, C.G.: Studies in the Logic of Confirmation. Mind 54, Part I, 1–26 (1945)

[R14.48] Hempel, C.G.: The Theoretician's Dilemma. In: Feigl, H., Scriven, M. (eds.) Minnesota Studies in the Philosophy of Science, vol. II, pp. 37–98 (1958)

[R14.49] Hempel, C.G.: Probabilistic Explanation. In: Brody, B.A. (ed.) Reading in the Philosophy of Science, pp. 28–38. Prentice-Hall, Englewood Cliffs (1970)

[R14.50] Hempel, C.G., Oppenheim, P.: Studies in the Logic of Explanation. Philosophy of Science 15, 135–175 (1948); also in Brody, B. A. (ed.) Reading in the Philosophy of Science, pp. 8–27. Prentice-Hall, Englewood Cliffs (1970)

[R14.51] Hempel, C.G., Oppenheim, P.: A Definition of Degree of Confirmation. Philosophy of Science 12, 98–115 (1945)

[R14.52] Heyting, A.: Intuitionism: An Introduction. North-Holland, Amsterdam (1971)

[R14.53] Hintikka, J. (ed.): The Philosophy of Mathematics. Oxford University Press, London (1969)

[R14.54] Hockney, D., et al. (eds.): Contemporary Research in Philosophical Logic and Linguistic Semantics. Reidel Pub., Co., Dordrecht-Holland (1975)

[R14.55] Hoyninggen-Huene, P., Wuketits, F.M. (eds.): Reductionism and Systems Theory in the Life Science: Some Problems and Perspectives. Kluwer Academic Pub., Dordrecht (1989)

[R14.56] Kedrov, B.M.: Toward the Methodological Analysis of Scientific Discovery. Soviet Studies in Philosophy 1, 45–65 (1962)

[R14.57] Kedrov, B.M.: On the Dialectics of Scientific Discovery. Soviet Studies in Philosophy 6, 16–27 (1967)

[R14.58] Kemeny, J.G., Oppenheim, P.: On Reduction. In: Brody, B.A. (ed.) Reading in the Philosophy of Science, pp. 307–318. Prentice-Hall, Englewood Cliffs (1970)

[R14.59] Klappholz, K.: Value Judgments of Economics. British Jour. of Philosophy 15, 97–114 (1964)

[R 14.60] Kleene, S.C.: On the Interpretation of Intuitionistic Number Theory. Journal of Symbolic Logic 10, 109–124 (1945)

[R14.61] Kmita, J.: The Methodology of Science as a Theoretical Discipline. Soviet Studies in Philosophy, 38–49 (Spring 1974)

[R14.62] Krupp, S.R. (ed.): The Structure of Economic Science. Prentice-Hall, Englewood Cliffs (1966)

[R14.63] Kuhn, T.: The Structure of Scientific Revolution. University of Chicago Press, Chicago (1970)

[R14.64] Kuhn, T.: The Function of Dogma in Scientific Research. In: Brody, B.A. (ed.) Reading in the Philosophy of Science, pp. 356–374. Prentice-Hall, Englewood Cliffs (1970)

[R14.65] Kuhn, T.: The Essential Tension: Selected Studies in Scientific Tradition and Change. University of Chicago Press, Chicago (1979)

[R14.66] Lakatos, I. (ed.): The Problem of Inductive Logic. North Holland, Amsterdam (1968)

[R14.67] Lakatos, I.: Proofs and Refutations: The Logic of Mathematical Discovery. Cambridge University Press, Cambridge (1976)

[R14.68] Lakatos, I.: Mathematics, Science and Epistemology: Philosophical Papers, vol. 2, Cambridge Univ. Press, Cambridge (1978); Worrall, J., Currie, G. (eds.)

[R14.69] Lakatos, I.: The Methodology of Scientific Research Programmes, vol. 1. Cambridge University Press, New York (1978)

[R14.70] Lakatos, I., Musgrave, A. (eds.): Criticism and the Growth of Knowledge, pp. 153–164. Cambridge University Press, New York (1979)

[R14.71] Lawson, T.: Economics and Reality. Routledge, New York (1977)

[R14.72] Lenzen, V.: Procedures of Empirical Science. In: Neurath, O., et al. (eds.) International Encyclopedia of Unified Science, vol. 1-10, pp. 280–338. University of Chicago Press, Chicago (1955)

[R14.73] Levi, I.: Must the Scientist make Value Judgments? In: Brody, B.A. (ed.) Reading in the Philosophy of Science, pp. 559–570. Prentice-Hall, Englewood Cliffs (1970)

[R14.74] Lewis, D.: Convention: A Philosophical Study. Harvard University Press, Cambridge (1969)

[R14.75] Mayer, T.: Truth versus Precision in Economics. Edward Elgar, London (1993)

[R14.76] Menger, C.: Investigations into the Method of the Social Sciences with Special Reference to Economics. New York University Press, New York (1985)

[R14.77] Mirowski, P. (ed.): The Reconstruction of Economic Theory. Kluwer Nijhoff, Boston (1986)

[R14.78] Mueller, I.: Philosophy of Mathematics and Deductive Structure in Euclid's Elements. MIT Press, Cambridge (1981)

[R14.79] Nagel, E.: Review: Karl Niebyl, Modern Mathematics and Some Problems of Quantity, Quality, and Motion in Economic Analysis. The Journal of Symbolic Logic, 74 (1940)

[R14.80] Nagel, E., et al. (eds.): Logic, Methodology, and the Philosophy of Science. Stanford University Press, Stanford (1962)

[R14.81] Narens, L.: A Theory of Belief for Scientific Refutations. Synthese 145, 397–423 (2005)

[R14.82] Niebyl, K.H.: Modern Mathematics and Some Problems of Quantity, Quality and Motion in Economic Analysis. Philosophy of Science 7(1), 103–120 (1940)

[R14.83] Neurath, O., et al. (eds.): International Encyclopedia of Unified Science, vol. 1-10. University of Chicago Press, Chicago (1955)

[R14.84] Neurath, O.: Unified Science as Encyclopedic. In: Neurath, O., et al. (eds.) International Encyclopedia of Unified Science, vol. 1-10, pp. 1–27. University of Chicago Press, Chicago (1955)

[R14.85] Nkrumah, K.: Consciencism. Heinemann, London (1964)

[R14.86] Planck, M.: The philosophy of Physics. Norton and Co., New York (1936)

[R14.87] Planck, M.: Scientific Autobiography and Other Papers. Greenwood, Westport (1971)

[R14.88] Planck, M.: The Meaning and Limits of Exact Science. In: Planck, M. (ed.) Scientific Autobiography and Other Papers, pp. 80–120. Greenwood, Westport (1971)

[R14.89] Polanyi, M.: Genius in Science. In: Cohen, R.S., Wartofsky, M.W. (eds.) Methodological and Historical Essays in the Natural and Social Sciences, pp. 57–71. D. Reidel Publishing Co., Dordrecht (1974)

[R14.90] Popper, K.: The Nature of Scientific Discovery. Harper and Row, New York (1968)

[R14.91] Putnam, H.: Models and Reality. In: Benecerraf, P., Putnam, H. (eds.) Philosophy of Mathematics: Selected Readings, pp. 421–444. Cambridge University Press, Cambridge (1983)

[R14.92] Reise, S.: The Universe of Meaning. The Philosophical Library, New York (1953)

[R14.93] Robinson, R.: Definition. Clarendon Press, Oxford (1950)

[R14.94] Rudner, R.: The Scientist qua Scientist Makes Value Judgments. Philosophy of Science 20, 1–6 (1953)

[R14.95] Russell, B.: Our Knowledge of the External World. Norton, New York (1929)

[R14.96] Russell, B.: Human Knowledge, Its Scope and Limits. Allen and Unwin, London (1948)

[R14.97] Russell, B.: Logic and Knowledge: Essays 1901-1950. Capricorn Books, New York (1971)

[R14.98] Russell, B.: An Inquiry into Meaning and Truth. Norton, New York (1940)

[R14.99] Russell, B.: Introduction to Mathematical Philosophy. George Allen and Unwin, London (1919)

[R14.100] Russell, B.: The Problems of Philosophy. Oxford University Press, Oxford (1978)

[R14.101] Ruzavin, G.I.: On the Problem of the Interrelations of Modern Formal Logic and Mathematical Logic. Soviet Studies in Philosophy 3(1), 34–44 (1964)

[R14.102] Scriven, M.: Explanations, Predictions, and Laws. In: Brody, B.A. (ed.) Reading in the Philosophy of Science, pp. 88–104. Prentice-Hall, Englewood Cliffs (1970)

[R14.103] Sellars, W.: The Language of Theories. In: Brody, B.A. (ed.) Reading in the Philosophy of Science, pp. 343–353. Prentice-Hall, Englewood Cliffs (1970)

[R14.104] Sterman, J.: The Growth of Knowledge: Testing a Theory of Scientific Revolutions with a Formal Model. Technological Forecasting and Social Change 28, 93–122 (1995)

[R14.105] Tullock, G.: The Organization of Inquiry. Liberty Fund Inc., Indianapolis (1966)

[R14.106] Van Fraassen, B.: Introduction to Philosophy of Space and Time. Random House, New York

[R14.107] Veldman, W.: A Survey of Intuitionistic Descriptive Set Theory. In: Petkov, P.P. (ed.) Mathematical Logic: Proceedings of the Heyting Conference, pp. 155–174. Plenum Press, New York (1990)

[R14.108] Vetrov, A.A.: Mathematical Logic and Modern Formal Logic. Soviet Studies in Philosophy 3(1), 24–33 (1964)

[R14.109] von Mises, L.: Epistemological Problems in Economics. New York University Press, New York (1981)

[R14.110] Wang, H.: Reflections on Kurt Gödel. MIT Press, Cambridge (1987)

[R14.111] Watkins, J.W.N.: The Paradoxes of Confirmation. In: Brody, B.A. (ed.) Reading in the Philosophy of Science, pp. 433–438. Prentice-Hall, Englewood Cliffs (1970)

[R14.112] Whitehead, A.N.: Process and Reality. The Free Press, New York (1978)

[R14.113] Wittgenstein, L.: Ttactatus Logico-philosophicus. The Humanities Press Inc., Atlantic Highlands (1974)

[R14.114] Woodger, J.H.: The Axiomatic Method in Biology. Cambridge University Press, Cambridge (1937)

[R14.115] Zeman, J.: Information, Knowledge and Time. In: Kubát, L., Zeman, J. (eds.) Entropy and Information. Elsevier, Amsterdam (1975)

R15. Riskiness, Decision-Choice Process and Paradoxes in Knowledge Constuction

[R15.1]　Allais, M.: The Foundations of the Theory of Utility and Risk: Some Central Points of the Discussions at the Oslo Conference. In: Hagen, O., Wenstøp, F. (eds.) Progess in Utility and Risk Theory, pp. 3–131. D. Reidel Pub., Dordrecht (1984)

[R15.2]　Allais, M., Hagen, O. (eds.): Expected Utility Hypotheses and the Allais Paradox. D. Reidel Pub., Dordrecht (1979)

[R15.3]　Anand, P.: Foundations of Rational Choice Under Risk. Oxford University Press, New York (1993)

[R15.4]　Anderson, N.H., Shanteau, J.C.: Information Integration in Risky Decision Making. Journal of Experimental Psychology 84, 441–451 (1970)

[R15.5]　Bar-Hillel, M., Margalit, A.: Newcombe's Paradox Revisited. British Journal of Philosophy of Science 23, 295–304 (1972)

[R15.6]　Campbell, R., Sowden, L. (eds.): Paradoxes of Rationality and Cooperation: Prisoner's Dilemma and Newcomb's Problem. Universith of British Columbia Press, Vancouver (1985)

[R15.7]　Crouch, E.A., et al.: Risk/Analysis. Ballinger, Cambridge (1982)

[R15.8]　Einhorn, H., Hogarth, R.M.: Ambiguity and Uncertainty in Probabilistic Inference. Psychological Review 92, 433–461 (1985)

[R15.9]　Ellsberg, D.: Risk, Ambiguity and the Savage Axioms. Quarterly Journal of Economics 75, 643–669 (1961)

[R15.10]　Friedman, M., Savage, L.J.: The Utility Analysis of Choice Involving Risk. Journal of Political Economy 56, 279–304

[R15.11]　Peter, G., Sahlin, N.: Unreliable Probabilities, Risk taking, and Decision Making. Synthese 53, 361–386 (1982)

[R15.12]　Handa, J.: Risk, Probability and a New Theory of Cardinal Utility. Journal of Political Economy 85, 97–122 (1977)

[R15.13]　Hurley, S.L.: Newcomb's Problem, Prisoner's Dilemma, and Collective Action. Synthese 86, 173–196 (1991)

[R15.14]　Harsanyi, J.C.: Cardinal Utility in Welfare Economics and in the Theory of Risk-Taking. Jour. Polit. Econ. 61, 434–435 (1953)

[R15.15]　Hart, A.G.: Risk, Uncertainty and Unprofitability of Compounding Probabilities. In: Lange, O., et al. (eds.) Mathematical Economics and Econometrics, pp. 110–118. Chicago University Press, Chicago (1942)

[R15.16]　Kahneman, D., Tversky, A.: Prospect Theory. Econometrica 47, 263–292 (1979)

[R15.17]　Karmarkar, U.S.: Subjectively Weighted Utility and Allais Paradox. Organization Behavior and Human Performance 24, 67–72 (1979)

[R15.18]　Kogan, N., Wallach, M.A.: Risk Taking: A Study in Cognition and Personality. Hold Rinehart and Winston, New York (1974)

[R15.19]　Levi, I.: Ignorance, Probability and Rational Choice. Synthese 53, 387–417 (1982)

[R15.20]　Levi, I.: Four Types of Ignorance. Social Research 44, 745–756 (1977)

[R15.21]　MacCrimmon, Larsson, S.: Utility Theory: Axioms Versus 'Paradoxes'. In: Allais, M., Hagen, O. (eds.) Expected Utility Hypotheses and the Allais Paradox, pp. 333–409. D. Reidel Pub., Dordrecht (1979)

[R15.22]　Priest, G.: Sorites and Identity. Logique et Analyse 34, 293–296 (1991)

[R15.23] Raiffa, H.: Risk, Ambiguity, and Savage Axioms: Comment. Quarterly Journal of Economics 77, 327–337 (1963)

[R15.24] Roberts, H.V.: Risk, Ambiguity, and Savage Axioms: Comment. Quarterly Journal of Economics 75, 690–695 (1961)

[R15.25] Sainsbury, R.M.: Paradoxes. Cambridge University Press, Cambridge (1995)

[R15.26] Savage, L.J.: The Foundations of Statistics. Wiley, New York (1954)

[R15.27] Shubik, M.: Information, Risk, Ignorance and Indeterminacy. Quarterly Journal of Economics 75, 643–669 (1961)

[R15.28] Simpson, P.B.: Risk Allowance for Price Expectation. Econometrica 18, 253–259 (1950)

[R15.29] Stigum, B.P., et al. (eds.): Foundations of Utility and Risk Theory with Applications. D. Reidel Publishing Com., Boston (1983)

[R15.30] Williamson, T.: Knowledge and its Limits. Oxford University Press, Oxford (2000)

[R15.31] Theil, H.: The Allocation of Power that Minimizes Tension. Operations Research 19, 977–982 (1971)

[R15.32] Theil, H.: On Estimation of Relationships Involving Qualitative Variables. American Journal of Sociology 76, 103–154 (1970)

R16. The Prescriptive Science, Theory of Planning and Cost-Benefit Analysis in Knowledge Construction

[R16.1] Alexander Ernest, R.: Approaches to Planning. Gordon and Breach, Philadelphia (1992)

[R16.2] Bailey, J.: Social Theory for Planning. Routledge and Kegan Paul, London (1975)

[R16.3] Burchell, R.W., Sternlieb, G. (eds.): Planning Theory in the 1980's: A Search for Future Directions. Rutgers University Center for Urban and Policy Research, New Brunswick (1978)

[R16.4] Camhis, M.: Planning Theory and Philosophy. Tavistock Publication, London (1979)

[R16.5] Chadwick, G.: A Systems View of Planning. Pergamon, Oxford (1971)

[R16.6] Cooke, P.: Theories of Planning and Special Development. Hutchinson, London (1983)

[R16.7] Dompere, K.K., Lawrence, T.: Planning. In: Hussain, S.B. (ed.) Encyclopedia of Capitalism, vol. II, pp. 649–653. Facts On File, Inc., New York (2004)

[R16.8] Dompere, K.K.: Cost-Benefit Analysis and the Theory of Fuzzy Decision: Identification and Measurement Theory. Springer, Heidelberg (2004)

[R16.9] Dompere, K.K.: Cost-Benefit Analysis and the Theory of Fuzzy Decision: Fuzzy Value Theory. Springer, Heidelberg (2004)

[R16.10] Dompere, K.K.: Fuzziness and the Market Mockery of Democracy: The Political Economy of Rent-Seeking and Profit-Harvesting. A Working Monograph on political Economy II, Department of Economics, Howard University, Washington, D.C.

[R16.11] Dompere, K.K.: Social Goal-Objective Formation, Democracy and National Interest: Political Economy under Fuzzy Rationality. A Working Monograph on Political Economy I, Department of Economics, Howard University, Washington, D.C.

[R16.12] Faludi, A.: Planning Theory. Pergamon, Oxford (1973)

[R16.13] Faludi, A. (ed.): A Reader in Planning Theory. Pergamon, Oxford (1973)

[R16.14] Harwood, E.C. (ed.): Reconstruction of Economics, American Institute For Economic Research, Great Barrington, Mass. (1955); Also in Dewey, J., Bently, A.: Knowing and the known, p. 269. Beacon Press, Boston (1949)

[R16.15] Kickert, W.J.M.: Organization of Decision-Making a Systems-Theoretic Approach. North-Holland, New York (1980)

[R16.16] Knight, F.H.: Risk, Uncertainty and Profit. University of Chicago Press, Chicago (1971)

R17. Social Sciences, Mathematics and the Problems of Exact and Inexact Methods of Thought

[R17.1] Ackoff, R.L.: Scientific Methods: Optimizing Applied Research Decisions. John Wiley, New York (1962)

[R17.2] Angyal, A.: The Structure of Wholes. Philosophy of Sciences 6(1), 23–37 (1939)

[R17.3] Bahm, A.J.: Organicism: The Philosophy of Interdependence. International Philosophical Quarterly VII(2) (1967)

[R17.4] Bealer, G.: Quality and Concept. Clarendon Press, Oxford (1982)

[R17.5] Black, M.: Critical Thinking. Prentice-Hall, Englewood Cliffs (1952)

[R17.6] Brewer, M.B., Collins, B.E. (eds.): Scientific Inquiry and Social Sciences. Jossey-Bass Pub., San Francisco (1981)

[R17.7] Campbell, D.T.: On the Conflicts Between Biological and Social Evolution and Between Psychology and Moral Tradition. American Psychologist 30, 1103–1126 (1975)

[R17.8] Churchman, C.W., Ratoosh, P. (eds.): Measurement: Definitions and Theories. John Wiley, New York (1959)

[R17.9] Foley, D.: Problems versus Conflicts Economic Theory and Ideology. In: American Economic Association Papers and Proceedings, vol. 65, pp. 231–237 (1975)

[R17.10] Garfinkel, A.: Forms of Explanation: Structures of Inquiry in Social Science. Yale University Press, New Haven (1981)

[R17.11] Georgescu-Roegen, N.: Analytical Economics. Harvard University Press, Cambridge (1967)

[R17.12] Gilolispie, C.: The Edge of Objectivity. Princeton University Press, Princeton (1960)

[R17.13] Hayek, F.A.: The Counter-Revolution of Science. Free Press of Glencoe Inc., New York (1952)

[R17.14] Laudan, L.: Progress and Its Problems: Towards a Theory of Scientific Growth. University of California Press, Berkeley (1961)

[R17.15] Marx, K.: The Poverty of Philosophy. International Pub., New York (1971)

[R17.16] Phillips, D.C.: Holistic Thought in Social Sciences. Stanford University Press, Stanford (1976)

[R17.17] Popper, K.: Objective Knowledge. Oxford University Press, Oxford (1972)

[R17.18] Rashevsky, N.: Organismic Sets: Outline of a General Theory of Biological and Social Organism. General Systems XII, 21–28 (1967)

[R17.19] Roberts, B., Holdren, B.: Theory of Social Process. Iowa University Press, Ames (1972)

[R17.20] Rudner, R.S.: Philosophy of Social Sciences. Prentice Hall, Englewood Cliffs (1966)

[R17.21] Simon, H.A.: The Structure of Ill-Structured Problems. Artificial Intelligence 4, 181–201 (1973)

[R17.22] Toulmin, S.: Foresight and understanding: An Enquiry into the Aims of Science. Harper and Row, New York (1961)

[R17.23] Winch, P.: The Idea of a Social Science. Humanities Press, New York (1958)

R18. Theories of Utility, Expected Utility and Exact Problems of Exact Methods

[R18.1]   Allais, M., Hagen, O. (eds.): Expected Utility Hypothesis and Allias Paradox. D. Reidel Pub., Dordrecht (1979)

[R18.2]   Chipman, J.S.: Foundations of Utility. Econometrica 28(2), 193–224 (1960)

[R18.3]   Eatwell, J., et al. (eds.): Utility and Probability. Norton, New York (1990)

[R18.4]   Fishburn, P.C.: The Foundations of Expected Utility. D. Reidel, Dordrecht (1982)

[R18.5]   Fishburn, P.C.: Utility Theory for Decision Making. Wiley, New York (1970)

[R18.6]   Luce, R.D., Suppes, P.: Preferences, Utility, and Subjective Probabilities. In: Luce, D.R., et al. (eds.) Handbook of Mathematical Psychology, pp. 42–49. Wiley, New York (1965)

[R18.7]   MacCrimmon, K., Larsson, S.: Utility Theory Versus Paradoxes. In: Allais, M., Hagen, O. (eds.) Expected Utility Hypothesis and Allias Paradox. D. Reidel Pub., Dordrecht (1979)

[R18.8]   Samuelson, P.: Probability and Attempts to Measure Utility. Economic Review 1, 167–173 (1950)

[R18.9]   Samuelson, P.: Probability, Utility, and Independence Axiom. Econometrica 20, 670–678 (1952)

[R18.10] Schoemaker, P.: The Expected Utility Model: Its Variants, Purposes, Evidence and Limitations. Journal of Economic Literature 20, 529–563 (1982)

[R18.11] Stigler, G.J.: The Development of Utility Theory I. Journal of Political Economy 58, 307–327 (1958)

[R18.12] Stigler, G.J.: The Development of Utility Theory II. Journal of Political Economy 58, 373–396 (1958)

[R18.13] Suppes, P., Winet, M.: An Axiomatization of Utility based on the Notion of Utility Differences. Management Science 1, 259–270 (1955)

[R18.14] von Neumann, J., Morgenstern, O.: Theory of Games and Economic Behavior. Princeton University Press, Princeton (1947)

R19. Vagueness, Approximation and Reasoning in the Knowledge Construction

[R19.1]   Adams, E.W., Levine, H.F.: On the Uncertainties Transmitted from Premises to Conclusions in deductive Inferences. Synthese 30, 429–460 (1975)

[R19.2]   Arbib, M.A.: The Metaphorical Brain. McGraw-Hill, New York (1971)

[R19.3]   Bečvář, J.: Notes on Vagueness and Mathematics. In: Skala, H.J., et al. (eds.) Aspects of Vagueness, pp. 1–11. D. Reidel Co., Dordrecht (1984)

[R19.4]   Black, M.: Vagueness: An Exercise in Logical Analysis. Philosophy of Science 17, 141–164 (1970)

[R19.5]   Black, M.: Reasoning with Loose Concepts. Dialogue 2, 1–12 (1973)

[R19.6]   Black, M.: Language and Philosophy. Cornell University Press, Ithaca (1949)

[R19.7]   Black, M.: The Analysis of Rules. In: Black, M. (ed.) Models and Metaphors: Studies in Language and Philosophy, pp. 95–139 (1962)

[R19.8]   Black, M.: Models and Metaphors: Studies in Language and Philosophy. Cornell University Press, Ithaca (1962)

[R19.9]   Black, M.: Margins of Precision. Cornell University Press, Ithaca (1970)

[R19.10] Boolos, G.S., Jeffrey, R.C.: Computability and Logic. Combridge University Press, New York (1989)

[R19.11] Cohen, P.R.: Heuristic Reasoning about uncertainty: An Artificial Intelligent Approach. Pitman, Boston (1985)

[R19.12] Darmstadter, H.: Better Theories. Philosophy of Science 42, 20–27 (1972)

[R19.13] Davis, M.: Computability and Unsolvability. McGraw-Hill, New York (1958)

[R19.14] Dummett, M.: Wang's Paradox. Synthese 30, 301–324 (1975)

[R19.15] Dummett, M.: Truth and Other Enigmas. Harvard University Press, Cambridge (1978)

[R19.16] Endicott, T.: Vagueness in the Law. Oxford University Press, Oxford (2000)

[R19.17] Evans, G.: Can there be Vague Objects? Analysis 38, 208 (1978)

[R19.18] Fine, K.: Vagueness, Truth and Logic. Synthese 54, 235–259 (1975)

[R19.19] Gale, S.: Inexactness, Fuzzy Sets and the Foundation of Behavioral Geography. Geographical Analysis 4(4), 337–349 (1972)

[R19.20] Ginsberg, M.L. (ed.): Readings in Non-monotonic Reason. Morgan Kaufman, Los Altos (1987)

[R19.21] Goguen, J.A.: The Logic of Inexact Concepts. Synthese 19, 325–373 (1968/1969)

[R19.22] Grafe, W.: Differences in Individuation and Vagueness. In: Hartkamper, A., Schmidt, H.-J. (eds.) Structure and Approximation in Physical Theories, pp. 113–122. Plenum Press, New York (1981)

[R19.23] Goguen, J.A.: The Logic of Inexact Concepts. Synthese 19 (1968-1969)

[R19.24] Graff, D., Timothy (eds.): Vagueness. Ashgate Publishing, Aldershot (2002)

[R19.25] Hartkämper, A., Schmidt, H.J. (eds.): Structure and Approximation in Physical Theories. Plenum Press, New York (1981)

[R19.26] Hersh, H.M., et al.: A Fuzzy Set Approach to Modifiers and Vagueness in Natural Language. J. Experimental 105, 254–276 (1976)

[R19.27] Hilpinen, R.: Approximate Truth and Truthlikeness. In: Pprelecki, M., et al. (eds.) Formal Methods in the Methodology of Empirical Sciences, pp. 19–42. Reidel, Ossolineum, Dordrecht, Wroclaw (1976)

[R19.28] Hockney, D., et al. (eds.): Contemporary Research in Philosophical Logic and Linguistic Semantics. Reidel Pub. Co., Dordrecht-Holland (1975)

[R19.29] Ulrich, H., et al. (eds.): Non-Clasical Logics and their Applications to Fuzzy Subsets: A Handbook of the Mathematical Foundations of Fuzzy Set Theory. Kluwer, Boston (1995)

[R19.30] Katz, M.: Inexact Geometry. Notre-Dame Journal of Formal Logic 21, 521–535 (1980)

[R19.31] Katz, M.: Measures of Proximity and Dominance. In: Proceedings of the Second
          World Conference on Mathematics at the Service of Man, pp. 370–377.
          Universidad Politecnica de Las Palmas (1982)
[R19.32] Katz, M.: The Logic of Approximation in Quantum Theory. Journal of
          Philosophical Logic 11, 215–228 (1982)
[R19.33] Keefe, R.: Theories of Vagueness. Cambridge University Press, Cambridge
          (2000)
[R19.34] Keefe, R., Smith, P. (eds.): Vagueness: A Reader. MIT Press, Cambridge (1996)
[R19.35] Kling, R.: Fuzzy Planner: Reasoning with Inexact Concepts in a Procedural
          Problem-solving Language. Jour. Cybernetics 3, 1–16 (1973)
[R19.36] Kruse, R.E., et al.: Uncertainty and Vagueness in Knowledge Based Systems:
          Numerical Methods. Springer, New York (1991)
[R19.37] Ludwig, G.: Imprecision in Physics. In: Hartkämper, A., Schmidt, H.J. (eds.)
          Structure and Approximation in Physical Theories, pp. 7–19. Plenum Press, New
          York (1981)
[R19.38] Kullback, S., Leibler, R.A.: Information and Sufficiency. Annals of Math.
          Statistics 22, 79–86 (1951)
[R19.39] Lakoff, G.: Hedges: A Study in Meaning Criteria and Logic of Fuzzy Concepts.
          In: Hockney, D., et al. (eds.) Contemporary Research in Philosophical Logic and
          Linguistic Semantics, pp. 221–271. Reidel Pub. Co., Dordrecht-Holland (1975)
[R19.40] Lakoff, G.: Hedges: A Study in Meaning Criteria and the Logic of Fuzzy
          Concepts. Jour. Philos. Logic 2, 458–508 (1973)
[R19.41] Levi, I.: The Enterprise of Knowledge. MIT Press, Cambridge (1980)
[R19.42] Łucasiewicz, J.: Selected Works: Studies in the Logical Foundations of
          Mathematics. North-Holland, Amsterdam (1970)
[R19.43] Machina, K.F.: Truth, Belief and Vagueness. Jour. Philos. Logic 5, 47–77 (1976)
[R19.44] Menges, G., et al.: On the Problem of Vagueness in the Social Sciences. In:
          Menges, G. (ed.) Information, Inference and Decision, pp. 51–61. D. Reidel
          Pub., Dordrecht (1974)
[R19.45] Merricks, T.: Varieties of Vagueness. Philosophy and Phenomenological
          Research 53, 145–157 (2001)
[R19.46] Mycielski, J.: On the Axiom of Determinateness. Fund. Mathematics 53, 205–
          224 (1964)
[R19.47] Mycielski, J.: On the Axiom of Determinateness II. Fund. Mathematics 59, 203–
          212 (1966)
[R19.48] Naess, A.: Towards a Theory of Interpretation and Preciseness. In: Linsky, L.
          (ed.) Semantics and the Philosophy of Language. Univ. of Illinois Press, Urbana
          (1951)
[R19.49] Narens, L.: The Theory of Belief. Journal of Mathematical Psychology 49, 1–31
          (2003)
[R19.50] Narens, L.: A Theory of Belief for Scientific Refutations. Synthese 145, 397–
          423 (2005)
[R19.51] Netto, A.B.: Fuzzy Classes. Notices, Amar., Math. Society 68T- H28, 945
          (1968)
[R19.52] Neurath, O., et al. (eds.): International Encyclopedia of Unified Science, vol. 1-
          10. University of Chicago Press, Chicago (1955)
[R19.53] Niebyl, K.H.: Modern Mathematics and Some Problems of Quantity, Quality
          and Motion in Economic Analysis. Science 7(1), 103–120 (1940)

[R19.54] Orlowska, E.: Representation of Vague Information. Information Systems 13(2), 167–174 (1988)

[R19.55] Parrat, L.G.: Probability and Experimental Errors in Science. John Wiley and Sons, New York (1961)

[R19.56] Raffman, D.: Vagueness and Context-sensitivity. Philosophical Studies 81, 175–192 (1996)

[R19.57] Reiss, S.: The Universe of Meaning. The Philosophical Library, New York (1953)

[R19.58] Russell, B.: Vagueness. Australian Journal of Philosophy 1, 84–92 (1923)

[R19.59] Russell, B.: An Inquiry into Meaning and Truth. Norton, New York (1940)

[R19.60] Shapiro, S.: Vagueness in Context. Oxford University Press, Oxford (2006)

[R19.61] Skala, H.J.: Modelling Vagueness. In: Gupta, M.M., Sanchez, E. (eds.) Fuzzy Information and Decision Processes, pp. 101–109. North-Holland, Amsterdam (1982)

[R19.62] Skala, H.J., et al. (eds.): Aspects of Vagueness. D. Reidel Co., Dordrecht (1984)

[R19.63] Sorensen, R.: Vagueness and Contradiction. Oxford University Press, Oxford (2001)

[R19.64] Tamburrini, G., Termini, S.: Some Foundational Problems in Formalization of Vagueness. In: Gupta, M.M., et al. (eds.) Fuzzy Information and Decision Processes, pp. 161–166. North Holland, Amsterdam (1982)

[R19.65] Termini, S.: Aspects of Vagueness and Some Epistemological Problems Related to their Formalization. In: Skala, H.J., et al. (eds.) Aspects of Vagueness, pp. 205–230. D. Reidel Co., Dordrecht (1984)

[R19.66] Tikhonov, A.N., Arsenin, V.Y.: Solutions of Ill-Posed Problems. John Wiley and Sons, New York (1977)

[R19.67] Tversky, A., Kahneman, D.: Judgments under Uncertainty: Heuristics and Biases. Science 185, 1124–1131 (1974)

[R19.68] Ursul, A.D.: The Problem of the Objectivity of Information. In: Kubát, L., Zeman, J. (eds.) Entropy and Information, pp. 187–230. Elsevier, Amsterdam (1975)

[R19.69] Vardi, M. (ed.): Proceedings of Second Conference on Theoretical Aspects of Reasoning about Knowledge. Morgan Kaufman, Asiloman (1988)

[R19.70] Verma, R.R.: Vagueness and the Principle of the Excluded Middle. Mind 79, 66–77 (1970)

[R19.71] Vetrov, A.A.: Mathematical Logic and Modern Formal Logic. Soviet Studies in Philosophy 3(1), 24–33 (1964)

[R19.72] von Mises, R.: Probability, Statistics and Truth. Dover Pub., New York (1981)

[R19.73] Williamson, T.: Vagueness. Routledge, London (1994)

[R19.74] Wiredu, J.E.: Truth as a Logical Constant With an Application to the Principle of the Excluded Millde. Philos. Quart. 25, 305–317 (1975)

[R19.75] Wright, C.: On Coherence of Vague Predicates. Synthese 30, 325–365 (1975)

[R19.76] Wright, C.: The Epistemic Conception of Vagueness. Southern Journal of Philosophy 33(suppl.), 133–159 (1995)

[R19.77] Zadeh, L.A.: A Theory of Commonsense Knowledge. In: Skala, H.J., et al. (eds.) Aspects of Vagueness, pp. 257–295. D. Reidel Co., Dordrecht (1984)

[R19.78] Zadeh, L.A.: The Concept of Linguistic Variable and its Application to Approximate reasoning. Information Science 8, 199–249 (1975); also in 9, 40–80

Index